Fundamentals of Supramolecular Chirality

Fundamentals of Supramolecular Chirality

Editors

Roberto Purrello • Alessandro D'Urso
University of Catania, Italy

W⊖ World Scientific

NEW JERSEY · LONDON · SINGAPORE · BEIJING · SHANGHAI · HONG KONG · TAIPEI · CHENNAI · TOKYO

Published by

World Scientific Publishing Europe Ltd.

57 Shelton Street, Covent Garden, London WC2H 9HE

Head office: 5 Toh Tuck Link, Singapore 596224

USA office: 27 Warren Street, Suite 401-402, Hackensack, NJ 07601

Library of Congress Cataloging-in-Publication Data

Names: Purrello, Roberto, editor. | D'Urso, Alessandro, editor.

Title: Fundamentals of supramolecular chirality / editors,
   Roberto Purrello, Alessandro D'Urso, University of Catania, Italy.

Description: New Jersey : World Scientific, [2022] | Includes bibliographical references and index.

Identifiers: LCCN 2021021094 | ISBN 9781800610248 (hardcover) |
   ISBN 9781800610255 (ebook) | ISBN 9781800610262 (ebook other)

Subjects: LCSH: Chirality. | Supramolecular chemistry.

Classification: LCC QD471 .F768 2022 | DDC 547/.1226--dc23

LC record available at https://lccn.loc.gov/2021021094

British Library Cataloguing-in-Publication Data

A catalogue record for this book is available from the British Library.

For any available supplementary material, please visit
https://www.worldscientific.com/worldscibooks/10.1142/Q0301#t=suppl

Desk Editors: Jayanthi Muthuswamy/Michael Beale/Shi Ying Koe

Typeset by Stallion Press
Email: enquiries@stallionpress.com

https://doi.org/10.1142/9781800610255_fmatter

About the Editors

Roberto Purrello is Full Professor of Chemistry at the Department of Chemical Sciences of the University of Catania. He received his Laurea Degree in 1979 (summa cum laude) and in 1983, he started his academic career in Catania as Assistant Professor. In 1988, he joined Tom Spiro's group at the Department of Chemistry of Princeton University to work on UV-resonance Raman of nucleic acids and proteins. From 1989 to 1990, he was appointed as Research Assistant in Princeton and in 1991 he was appointed as Associate Professor in Catania. In 1992, he visited the Department of Chemistry of the Swarthmore College (PA, USA) starting a collaboration with Prof. R. F. Pasternack. In 1999, he was appointed as Visiting Fellow at the Department of Chemistry in Davis (CA, USA) in the laboratory of Prof. K. M. Smith. Since 2003, he has been collaborating with Prof. N. Berova at the Department of Chemistry of Columbia University (NY, USA). He is member of the Editorial Board of *Chirality* from July 2010.

Alessandro D'Urso is Associate Professor at the Department of Chemical Sciences of the University of Catania. His research activity is mainly focused on interactions of water-soluble porphyrinoids with biomolecules, as chiroptical reporters of conformations. He got his PhD in Chemistry in 2010 at the University of Catania. In 2009, he was a Visiting Scholar at the University of Wyoming (Supervisor: Prof. Milan Balaz) working on chiroptical sensors for Z-DNA. In 2010, he was invited as a Research Assistant at the Chemistry Department of Doane College (Prof. Andrea Holmes). Then in 2012, he was a Postdoctoral Research Scientist at Columbia University, Chemistry Department in the lab of Prof. Koji Nakanishi and Prof. Nina Berova. He has been awarded with "SPP/JPP Young Investigator Award 2016" and with "Best Poster Award" at 14th International Conference on Chiroptical Spectroscopy 2013 in Nashville, USA. He has authored over 60 papers published in international peer-reviewed journals and 5 chapters in international books and has more than 1000 citations.

Contents

Introduction — Chirality: What Is It, What Is It For?

Roberto Purrello

Department of Chemical Sciences, University of Catania, Viale A. Doria 6, Catania 95125, Italy

rpurrello@unict.it

Chirality, defined as "the property of an object of not being superimposable with its mirror image", is widespread in the universe where it is expressed at different levels: from subatomic to galactic. In (bio) chemistry, chirality has two relevant levels: the molecular and the supramolecular ones. Both these levels have specific meanings, and biological systems are naturally organized in such a way that they may recognize and differently interact with molecules, which are identical in all physico-chemical properties but those deriving from their (chiral) three-dimensional shape.

Involved on the Earth since the origin of life and fully exploited along the way of evolution, chirality has a plethora of roles in biological systems, from discrimination to regulation. The molecular bricks of life — amino acids and nucleotides — are chiral, as are their polymeric forms. However — and very interestingly — the different molecularity of the chiral species means quite different roles in nature. The differences between these two levels are exemplified and actualized in nature, where (bio)chemical chirality finds a very effective representation of its functional and sophisticated relevance.

Molecular chirality of biological molecules is, in fact, mainly related to the fixed disposition of four different groups around a stereogenic central atom. This peculiarity is fundamental to carry specific information (L- and not D-amino acids are absorbed by our organism, the smell and taste of L- and D- molecules are different) but limits their biological role. On the contrary, chiral biopolymers — which have complex spatial relationships between the monomeric constituents — are flexible and their conformational chirality can be modulated with deep consequences on their biological meaning.

Chirality, at both molecular and supramolecular levels, is of paramount importance for our daily life. For example, the interaction of the two enantiomeric forms with their biological target/receptor is differently transduced and can lead to different, and sometimes unwanted, effects. For example, L- and D-carvone are the flavor of mint and cumin, respectively. More importantly, also enantiomeric forms of medicines can behave very differently with their target in the organism causing unhealthy effects, like the cases of thalidomide. Finally, chirality is relevant to design materials that might find applications in functional soft matter, smart materials, and other fascinating technological fields.

Yet, as already observed for the biological level of chirality, the study of the expression of chirality at a supramolecular level is particularly intriguing. At this level, chirality may also arise from non-covalent organization, and even more intriguingly, the single molecular components of the supramolecular chiral assemblies might not be chiral themselves. Chiral non-covalent assemblies formed from non-chiral molecular entities are, from many respects, the most intriguing to study because they allow for an almost endless choice of molecules to be assembled in a chiral fashion, for a wide range of applications. Another interesting point of the chirality of non-covalent supramolecular assemblies is that their chirality will be also expressed at the mesoscopic level: the same amplification mechanism that leads from the nano- to the micro-level will amplify their own chirality to the meso-level paving the way to the synthesis of new materials.

In this book, we focus on several aspects of supramolecular chirality analyzed in five chapters. The induction of chirality in supramolecular system has to follow precise hierarchical roles; indeed, in Chapter 1, several examples of supramolecular chiral systems were reported highlighting different instances where clear pathways for chiral induction can be deduced, even if these routes are difficult to predict. In particular, the structures and properties of some selected examples of supramolecular chirality in natural compounds were analyzed in detail in Chapter 2. Even the fundamental strategies and challenges for design and construction of supramolecular chirogenic systems, and the major chirogenic applications like chiral induction, sensing/recognition, discrimination, and photochiro-genesis processes on the origin of supramolecular chirality was reviewed in Chapter 3. Finally in the last two chapters two amazing phenomena are presented, memorization of chirality and effect of physical forces on supramolecular chirality.

Chapter 1

Hierarchies in the Transfer of Molecular to Supramolecular Chirality

David B. Amabilino[*], Glenieliz C. Dizon[†],
C. Elizabeth Killalea[‡], and Ajith R. Mallia[§]

*The GSK Carbon Neutral Laboratories for Sustainable Chemistry,
School of Chemistry, University of Nottingham,
Triumph Road, Nottingham NG7 2TU, UK*

[*]*david.amabilino@nottingham.ac.uk*
[†]*Glenieliz.Dizon@nottingham.ac.uk*
[‡]*Catherine.Killalea@nottingham.ac.uk*
[§]*ajithrmallia@gmail.com*

The structural pathways of chirality transfer from individual stereogenic moieties in molecules to supramolecular assemblies and materials is discussed. It is shown that there are instances where clear pathways for chiral induction can be deduced, even if these routes are difficult to predict. The methods to observe these hierarchies, at different levels of dimensionality, from fibers to films and macroscopic objects like crystals, are presented, and the consequences of the chiral transfer are shown to augur well, leading to some fascinating properties.

1. Preamble

The preparation of the synthetic materials described in this contribution are often inspired by our surroundings, and while manmade objects are often achiral, asymmetry is inherent to the natural world,[1] from galaxies[2] to the tiniest crystals.[3,4] The most obvious manifestations of single-handed chirality around us are in the plants that twist around themselves to form vines, the shells of snails on land or sea, and the seeds of certain trees (ash and sycamore for example) that spin as they fall. All of these macroscopic objects are, at least partially, a manifestation of the handedness of the materials of which they are comprised. And yet, the generality of the structural pathways they take toward single handedness are largely unknown and unpredictable. In these biological systems, chirality must be transferred from the natural polymers (based on L amino acids and D sugars, that are themselves made by the larger structures of course) up through various hierarchical states to the macroscopic object (Figure 1).[5,6] The same is true for chiral synthetic systems, which can display chiral morphology with enantiomorphism and/or chiral effects as materials as a result of the influence of molecular-level stereogenic features.[7-10] The hierarchical transfer of chirality is a result of self-assembly and applies to all kinds of materials, from crystals to cells.[11]

There are some remarkable outcomes in the formation of synthetic supramolecular systems where molecular chirality is transferred up to supramolecular structure[12] and beyond.[13] The most obvious signature of chirality to our perception is the helix: Particularly, helical systems based on small molecules or polymers have demonstrated hierarchical chirality transfer.[14] But other shapes and complicated structures arise, as seen in

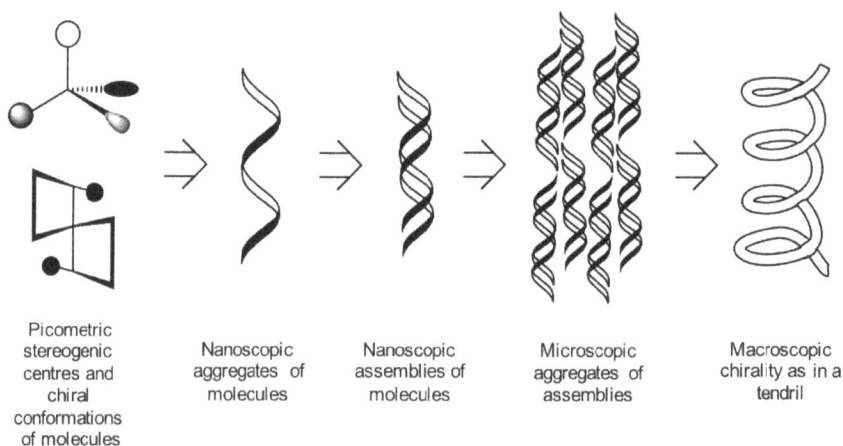

| Picometric stereogenic centres and chiral conformations of molecules | Nanoscopic aggregates of molecules | Nanoscopic assemblies of molecules | Microscopic aggregates of assemblies | Macroscopic chirality as in a tendril |

Figure 1.   A scale of hierarchies for chiral chemical objects.

virtually any natural object whose irregular structure is difficult to qualify in a chiral sense. At present, there is no universal model for how chirality is transferred, it is system dependent and scale dependent. This situation results from the very different forces happening at the molecular level in the nanometer region. There are a number of mechanisms by which molecular-level chirality is passed up a hierarchy, as shall be discussed in the next section. Subsequently, examples of chiral transfer in systems with different dimensionalities will be shown before we present our personal outlook for understanding and using the phenomena.

2. Chirality and Its Transfer

The three sources of structural chirality in molecular chemistry — stereogenic centers (point chirality), axes of chirality, or planar chirality — can all have an influence on hierarchical assembly.[15] The vast majority of studies on chirality transfer employing chiral compounds involve the use of stereogenic centers at some point of the molecule, and of course, natural systems all involve this source of asymmetry. But chiral transfer can also take place in apparently achiral systems with none of these sources. It is only at the supramolecular level that the chirality evolves, and spontaneous symmetry breaking[16] occurs, mainly, if not always, leading to spontaneous resolution.[17,18]

2.1. *Mechanisms of chiral transfer*

The more common sources of chirality in a molecule, be they stereogenic centers at atoms or axial chirality elements in atropoisomeric compounds, are defined by their connectivity and arrangement in three-dimensional space. It is the fit of neighboring molecules together, the contact between them, that determines the transfer of chirality. The presence of the stereogenic center favors one of the multiple diastereoisomeric outcomes of forming a supramolecular dimer (Figure 2) in the first case, and the subsequent oligomers should further aggregation take place. Similarly, chiral conformations around rigid segments of the molecular structures giving rise to atropoisomeric structures, or transiently chiral shapes can assemble to give dimers that are twisted in a specific way to one another. However, it is clear that if both enantiomers exist in solution, those molecules could assemble as a homochiral or heterochiral pair. These situations are

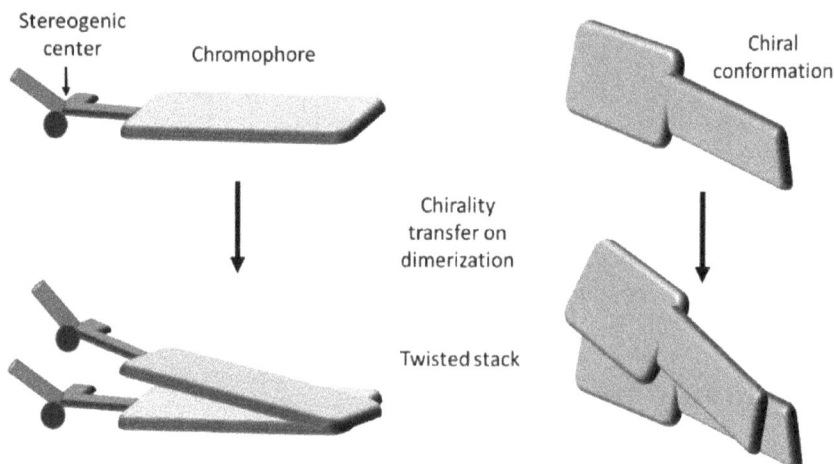

Figure 2.   Cartoons showing hypothetical chirality transfer from (on the left) a stereogenic center to the arrangement of two flat and long chromophores in a dimer supramolecular object and (on the right) a chiral conformation to a twisted stack. The stereogenic center can have very weak optical activity, but the dimer can display very strong chiroptical effects depending on the proximity and arrangement of the chromophores, and assemblies of twisted molecules can affect optical activity dramatically.

observed and led to the idea of enantiophilic and enantiophobic molecules, where the latter would lead to homochiral spontaneously resolved assemblies.[18]

The interactions between two molecules as the start of a hierarchical self-assembly process could be propagated by, and could even incorporate, non-identical molecules in the aggregation that are still able to pass on the initial chirality. In this way, chirality can be transferred and "amplified".[19,20] Two of the most important concepts in the amplification of transfer of chirality of growing supramolecular structures are "majority rules" and "sergeants and soldiers", which are in effect an amplification of the local stereochemistry of one of the components (Figure 3).

An early example of the sergeants and soldiers approach applied to a supramolecular system was the formation of stacked hydrogen-bonded assemblies based on dimelamine-cyanurate derivative rosettes.[21] The system operates under thermodynamic control where the kinetics of exchange of the components is different, the dimelamine derivative exchanges slowly and on a rate similar to inversion of helicity of the stacks.

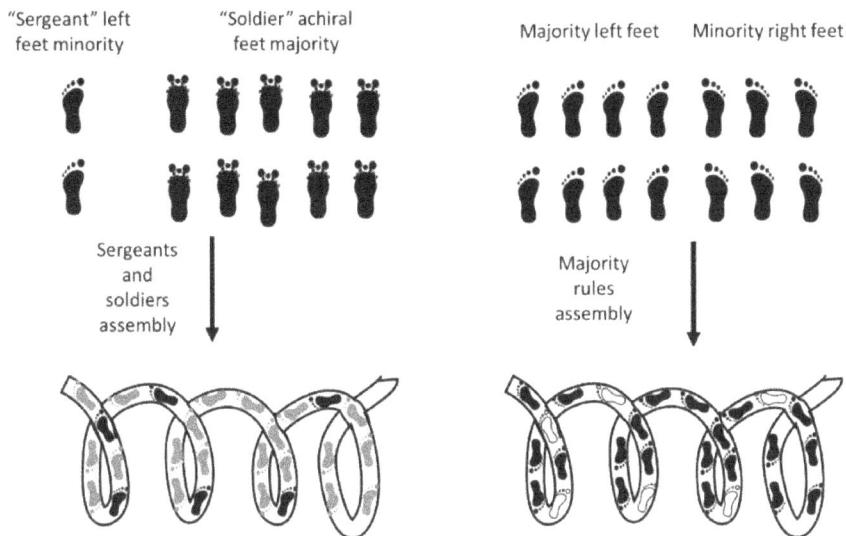

Figure 3. Representations of the sergeants and soldiers and majority rules effects. In the sergeants and soldiers type system, a minority chiral compound — the sergeant — assembles with a related achiral but prochiral molecule — the soldier — that is made to twist in the way that is thermodynamically preferred by the former, as indicated by the grey foot in the assembly. In the majority rules system, a mixture of enantiomers with an excess of one mirror image assembles such that the molecule in excess, in the above case the left foot, is able to impose its chirality in the assembly on the right foot, shown in the self-assembled structure as a colorless foot, as the diastereomeric conformation is not the lowest energy state of the molecule, although the system is at a thermodynamic minimum.

Therefore, in this case, the chiral transmission was relatively poor. On the other hand, when the stereogenic center was incorporated into the cyanurate derivative, which exchanges quickly compared to the rate of rosette chirality inversion, the amplification of chirality was much more effective.

Within the area of chirality transfer from stereogenic centers through the interaction of neighboring homochiral molecules or through the amplification of chirality by the sergeants and soldiers or majority rules mechanisms, the position of the point chirality determines the effectiveness and handedness of the induction of the chiral structure. This general phenomenon, termed the odd–even effect in its simplest form, was probably first uncovered in studies on achiral liquid crystals[22] and subsequently in the induction of cholesteric phases in achiral nematic phases using chiral dopants[23] as well as other chiral mesophases.[24,25]

There are two main types of structural odd–even effect related with the hierarchical transfer of chirality. In one, the position of the stereogenic center along an alkyl chain can be changed, leading to opposing chirality of the induced structures.[26] The reason for this alternance is that (as indicated in Figure 4) the substituent to the chain is directed to an opposite face in odd or even cases, and also the dipole moment associated with this substituent is changed, both factors determining the chirality of the assembled molecules.[27] In the other type of structural change, the length of the achiral methylene linkage between two moieties is modified,[28] and in this case the terminal groups end up being oriented in different ways (Figure 4). For example, the peptides shown in Figure 4 comprise identical amino acid residue sequences spaced either side of four different aromatic moieties by aliphatic chains of two odd and two even number chains.[29] The aggregates of these compounds (made by acidifying a dissolved state of the peptides in basic water) show an odd–even effect for the compounds with thiophene and benzene ring spacers, but not for the other two units. These results led to the conclusion that hydrogen-bonding and hydrophobically favored packing were the main forces determining whether the structure would affect chirality or not, a hypothesis supported

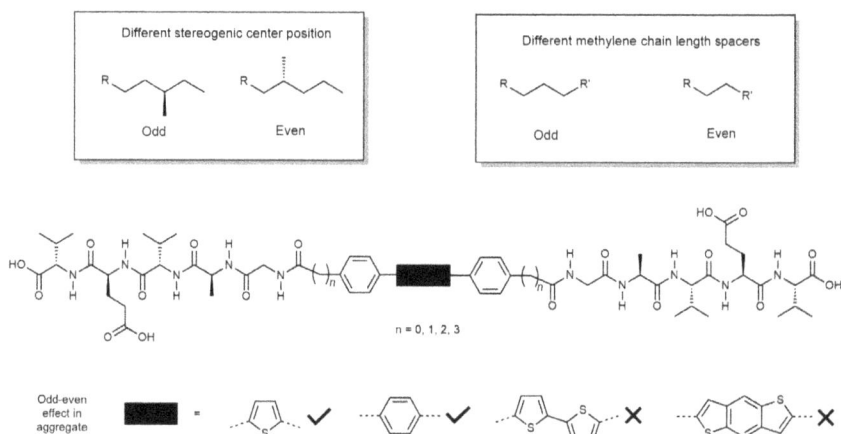

Figure 4.   A classification of structures that show odd–even effects where the chirality of superstructures depend on the position of a stereogenic center in an alkyl chain and the length of a methylene carbon linkage between two units that could be chiral. The structure of a bis-peptide[29] where the modification of the aromatic spacer leads to odd–even effects in the assembly (indicated by the tick) or no odd–even effect (indicated by the cross).

by molecular mechanics simulations in concert with machine learning analysis methods. This combination of theoretical and experimental information is surely a good example of how to learn more from assembly at this relatively simple one-dimensional aggregate level and beyond.

Helices can template transfer of chirality to guests, inducing chiral conformations. An example of this effect is the helical conformation of oligosilanes when they are bound inside the single helices of carboxymethylated amylose or the double helices formed by schizophyllan, both in water (Figure 5).[30] The functionalized amylose forms a complex with the oligosilane in water that can be isolated and partially redissolved for characterization, which leads to the inference that the carbohydrate forms an 8_1 helix with a helical silane (forming a 15/7 helix) inside it. The helicity of the complex of the oligosilane inside schizophyllan is observed to be the opposite of the same compound inside the amylose derivative.

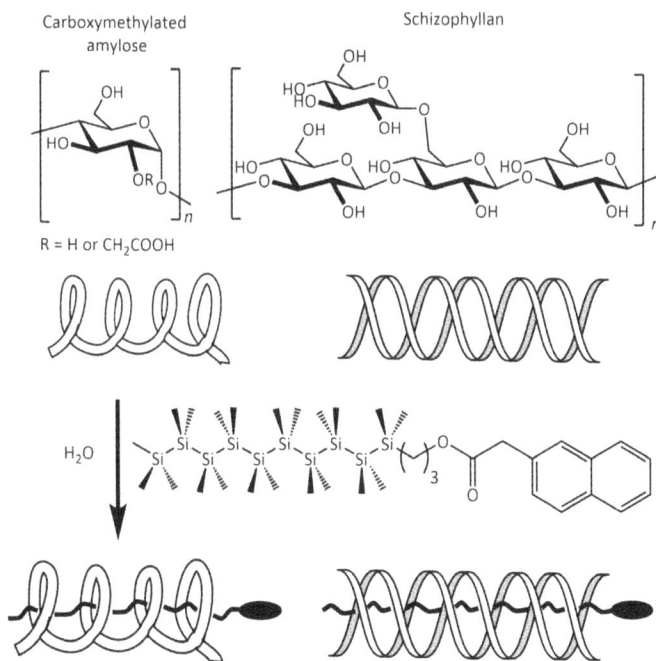

Figure 5.   The chemical structures of carboxymethylated amylose and schizophyllan and cartoons of the helices they form.[30] When the oligosilane shown (in cartoon form as a black wavy line and ellipse) is bound to these helices in water, the helicity of the helical host determines the chirality of the helix formed by the guest.

Therefore, the handedness of the surrounding helix — right in the case of schizophyllan, left in the case of carboxymethylated amylose — determines the conformation adopted by the oligosilane. This supramolecular wrapping strategy by helix-forming polysaccharides is a general strategy that can be used to transfer chirality.[31]

The reverse type of chiral induction is also possible, where a chiral guest induces chirality on an achiral molecule that can adopt a helical secondary structure. Foldamers are an excellent example of this effect.[32,33] These molecules often exist in dynamic equilibria between open forms and helical conformations. Binding of chiral guests can shift the equilibrium to a helix with a handedness that is determined by the chirality of the molecule included in the cylindrical cavity of the folded form, as shown by the example in Figure 6.[34] The achiral molecule undergoes reversible folding in polar solvents thanks to the rotational freedom around the alkyne linkage so that the aromatic moieties can stack intramolecularly, leaving a cavity in the center. When either enantiomer of α-pinene is introduced to a solution of the foldamer, a 1:1 complex is formed where each terpene induces an opposite preferred handedness to the helical host. Conceptually similar foldamers that incorporate recognition sites inside the cavity have led to remarkable guest binding, including the recent report of a double-helical foldamer that was able to bind heteromeric pairs

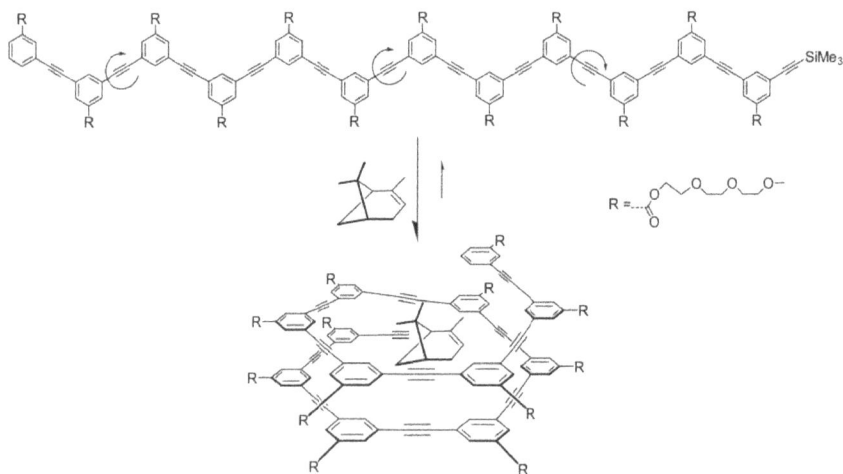

Figure 6.   A representation of a foldamer in its extended form and the helical conformation it adopts around the guest molecule α-pinene when the terpene is bound in solution.

of monosaccharides.[35] These complexes are usually studied in solution as zero dimensional and designed to crystallize; there is an example of an oligoamide foldamer that self-assembles into microfibers that can be twisted,[36] although the effects of chirality have yet to be explored, and achiral hydrogen-bonded hydrazide foldamers can complex sugars to give fibers that gelate organic solvents.[37] The chirality, while evident spectroscopically, could not be observed in the morphology of the fibers.

The mechanisms of chiral transfer above are at a low hierarchical level, but it is essential to understand these effects so that the next length scale can be understood. The way in which these aggregates themselves come together to form larger objects is presently much harder to explain in synthetic systems. There are cases of natural systems where structural chirality can be traced. Cellulose nanocrystals[38] is one such case where control of the arrangement gives tunable optical properties.[39] However, chirality does not necessarily transfer and chiral morphologies are not observed always: Very often, the molecular chirality is not expressed at all, but those results are probably largely unreported in the open literature and are not easy to track in a comprehensive manner because of the negative perceived nature of the results. Pasteur was brilliantly lucky in having the enantiomers of sodium ammonium tartrate as enantiomorphous crystals,[40–42] because most chiral crystals do not exhibit this behavior. Most crystals of enantiomeric materials appear completely achiral under microscopes. So, to explain why chirality transfers, we probably also have to explain why it does not! The challenge facing us is largely to do with observing and modeling structures over various length scales; so, now it is appropriate to consider how that is done.

2.2. *How do we see hierarchical chirality transfer?*

It depends! Chirality is most usually detected in molecules and their aggregates by optical activity (be it by absorbance of light, luminescence, or birefringence) and can indicate transfer of chirality from stereogenic centers to the arrangements of chromophores in space, as shown in the cartoon in Figure 2. In supramolecular systems with chromophores that form hierarchical structures, the technique of choice in their nascent form in solution is surely circular dichroism (CD) spectroscopy because of its sensitivity.[43,44] From its initial use in the characterization of secondary structures in proteins[45,46] or induced chirality in dyes,[47,48] where one of the earliest and nicest examples of induced optical activity was actually studied previously using optical rotatory dispersion in cationic dye-polyglutamate complexes,[49] CD is generally more intuitive because the bands correlate with specific optically induced transitions rather than optical rotation (although this parameter does change near absorption bands as well). CD spectroscopy is also useful when performed in the solid state.[50,51] Peptide secondary structure in the solid state can be probed using the technique,[52] and it can be used for all kinds of synthetic materials including thin films of π-functional compounds.[53]

Chiroptical effects in luminescence can also be studied and provides complementary information to the absorption spectra. Circularly polarized luminescence (CPL) shows asymmetry in aggregated systems and is especially useful for molecules demonstrating aggregation-induced emission (AIE). An early example in this burgeoning area was the aggregation of a molecule containing a luminogenic silole core linked through click chemistry to a sugar moiety whose aggregate shows greatly enhanced fluorescence compared to the monomer as well as circularly polarized emission.[54] This turned out to be a general phenomenon that has been observed in many self-assembling systems.[55,56]

Vibrational circular dichroism[57] (VCD) is a technique that is sensitive and specific to the environment of functional groups and can be used to probe self-assembly and hierarchical processes.[58] For example, the self-assembly of guanosine-5'-hydrazide in the presence of sodium cations leads to the formation of stacked quartets of the molecules that ultimately form fibers, and the process of disassociation of these supramolecular objects can be observed clearly with VCD.[59] Given the strength and sensitivity of the technique, the cases of its use for studying this kind of process are relatively rare.

Modeling of different kinds of chiral systems over many length scales is increasingly accessible and reliable.[60,61] The combination of modeling to determine tertiary structures and simulation of the aforementioned spectroscopic characteristics of the assemblies are powerful tools in understanding the organization of chiral systems.[62–65]

All those techniques largely look at microscopic chirality, but when hierarchical transfer takes place, microscopies are the techniques of choice. In a rough order of lateral resolution, these are scanning tunneling microscopy (STM, only for conducting surfaces), atomic force microscopy (AFM), transmission electron microscopy (TEM), scanning electron microscopy (SEM), and, of course, the gamut of optical microscopies. All these techniques lend themselves particularly well to the characterization of fibers,[66] and examples of them will be given later in this text.

3. Chirality Transfer Hierarchy in Single Organic Materials

In the following discussion, materials in which hierarchical transfer of chirality is observed are divided up into essentially one-dimensional systems where fibers form that can then intertwine, followed by layers and films of materials and finally three-dimensional crystals.

3.1. *From point chirality to single handed fibers and tubes*

Helical twisted fibers visible in an optical microscope were seen in 1945 from anteiso fatty acids (with an (*S*)-*sec*-butyl group at the alkyl termini) 24-methylhexacosanoic acid and 28-methyltriacontanoic acid, whereas other crystals of compounds in this series did not display twisting.[67] In 1965, enantiomorphism of lithium 12-hydroxystearate's helical fibers was demonstrated by studying the self-assembly of the pure enantiomers.[68] Several salts of the D enantiomer of 12-hydroxystearate showed right-handed helices in electron micrographs, with the size and pitch of the fibers dependent on exact composition, an exception being the calcium salt that showed a mixture of left- and right-handed fibers. The racemate showed achiral fibers, suggesting the formation of a true racemate in the material. Since then, single-handed fibers comprising a very wide range of organic and coordination compounds have been prepared and studied.[69] Some examples are shown here to demonstrate the scope and the frontiers of knowledge.

Gels are a particular area where fibers are usually intrinsic to their immobilization of solvents, especially low molecular weight gelators (LMWGs).[70] Gels are heterogeneous biphasic three-dimensional materials comprising a majority dispersed phase (fluid) within a minority continuous phase (solid), which in the case of LMWGs are usually thermally reversible (Figure 7). The design[71] and characterization[72] of chiral self-assembled fibrillar networks of many LMWGs has shown twisted, helical, or cylindrical tubular morphologies, depending on gelator constitution. Most supramolecular gel networks are formed through non-covalent forces[73] such as hydrogen-bonding, π–π stacking, van der Waals interactions, metal coordination, and dipole–dipole interactions *via* nucleation-controlled self-assembly of the gelator molecules.[74] The fibers are interesting for a number of materials areas including light harvesting.[75]

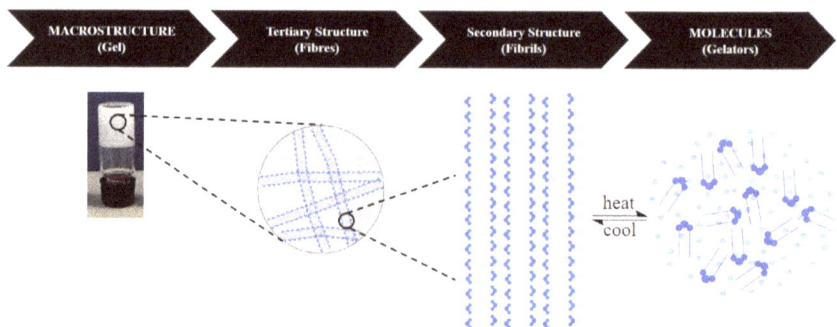

Figure 7.   The hierarchy of structure in a physical gel.

Stereochemistry can determine the arrangement of gelator molecules as well as the solvent (if the solvent is chiral) that may affect the formation of the network within the gel.[76] The physical or biological properties as well as the means in which the material reacts toward its chemical environment are affected by its stereochemistry. While chirality is not a requirement for a molecule to be an effective gelator (there are many achiral LMWGs), a remarkable number are chiral. Several chiral LMWGs immobilize water, giving hydrogels.[77] Recently, there is an increase of chiral gelators and, in most cases, the enantiomers of the chiral analogs are more effective gelators than the racemic mixtures.[13,78–80] The effects of chirality on molecular gel formation have been studied in detail, focusing on how self-assembly responds to mixtures of enantiomers as well as translating the molecular-scale chirality up to the nanoscale level[81] and even centimeters.[82] In analogy to the outcomes of three-dimensional crystallization, racemic mixtures of gelators can either co-assemble or self-assemble into fibers that are either (i) true racemates, (ii) pseudo racemates, or (iii) homochiral and self-sorted (Figure 8).

Like other soft materials, interpreting the morphologies of dried gels can cause difficulties in understanding gel structure because drying can affect the fiber networks significantly.[83] Although micrographs can give indications of the xerogel's nanostructures in different solvents, they are not always the most representative way of studying what is truly present within the solvated gel-phase. But differences can be seen between stereoisomers in xerogel form. The gelation behavior of a bis(urea) LMWG in its racemic form (mixture of enantiomers), mixture of diastereoisomers

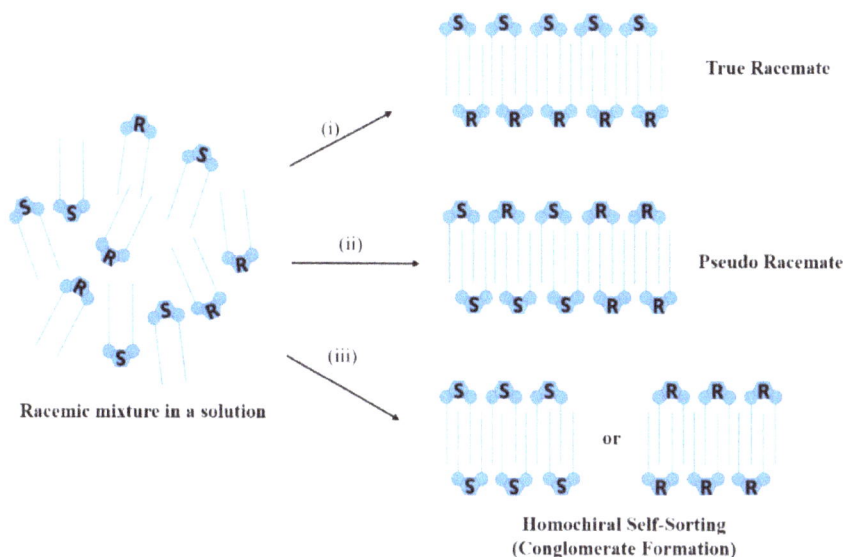

Figure 8. Outcomes of gel formation by a racemic compound or a mixture of chiral conformers.

(enantiomers and meso compound), and as its separate enantiomers showed better mechanical and thermal stability than the mixture of enantiomers, where fibers were found to exist as a racemate and areas of pure enantiomer.[84] Micrographs clearly indicate the difference in morphology of the different chiral materials. Helicity of single fibers is seen in the SEM micrograph of the enantiopure gelator, a feature attributed to the successful transfer of molecular chirality to the hierarchical aggregates (Figure 9(a)). On the other hand, tape-like fibers and no helical fibers, are seen in the xerogel of the mixture of diastereoisomers (Figure 9(b)).

An exciting area of research into hierarchical systems in general, and gels in particular, is that of responsive soft materials.[85,86] A molecule incorporating a four-state chiroptical switching cycle has been described showing transcription of chirality in a gel material (Figure 10).[87] The described cycle allows the photocontrol of chirality in the molecular as well as the supramolecular level *via* appended hydrogen-bonding units. According to the obtained TEM micrographs, gels of the open form (O) aggregate into one chiral conformation (P) leading to the formation of right-handed *P*-helical fibers. The photochemical ring closure in the gel

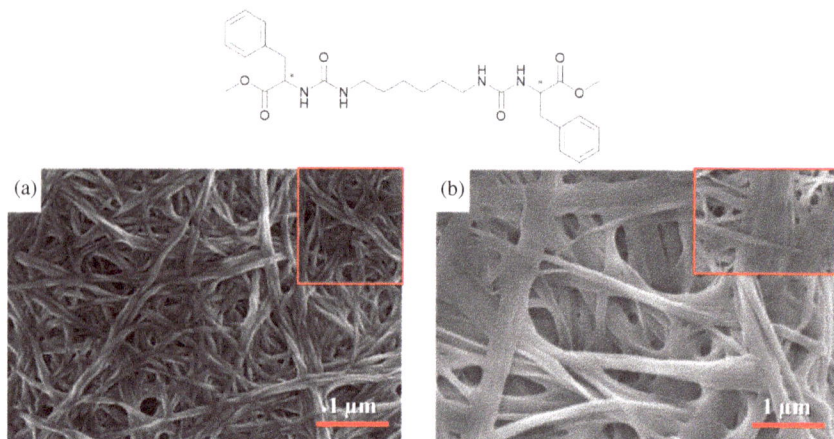

Figure 9.   SEM micrographs of (a) the (*S,S*) enantiomer and (b) the bis-urea prepared from racemic amine bis(urea) xerogels in ethyl acetate; inset shows the magnified images. Reprinted with permission from Ref. [84].

state maintains chirality to yield the closed form (C) with a high diastereomeric excess (de). The work demonstrates important design aspects, such as the molecular chirality of the materials. During aggregation or self-assembly, the chirality rules both the stability and the helicity. When the material is in solution, diastereoselectivity is not present; however, after aggregation of the material, nearly absolute stereocontrol is exerted at the molecular level.

More generally in nanofibers, C_3-symmetric molecular architectures have gained wide attention in forming well-defined nano-assemblies[19,88] where a central phenyl or cyclohexyl ring functionalized at the 1, 3, and 5 positions are often used as the core. Chiral functionalities are usually introduced into C_3-symmetric molecules *via* amide or urea bonds connected to the central tricarboxyl or triamine cores (Figure 11). As a consequence of (i) steric hindrance about the core and (ii) wedged substituents causing the rigidity through intramolecular hydrogen-bonding, the C_3-symmetric molecules often adopt a propeller-like conformation. During stacking, the chiral elements play a pivotal role in determining the chirality of the assemblies and in transferring the molecular chirality. The amide bonds at 1, 3, and 5 positions form intermolecular hydrogen bonds, which contribute to the one-dimensional growth of columnar-type supramolecular polymers.[89] The efficacy of the chirality transfer is largely

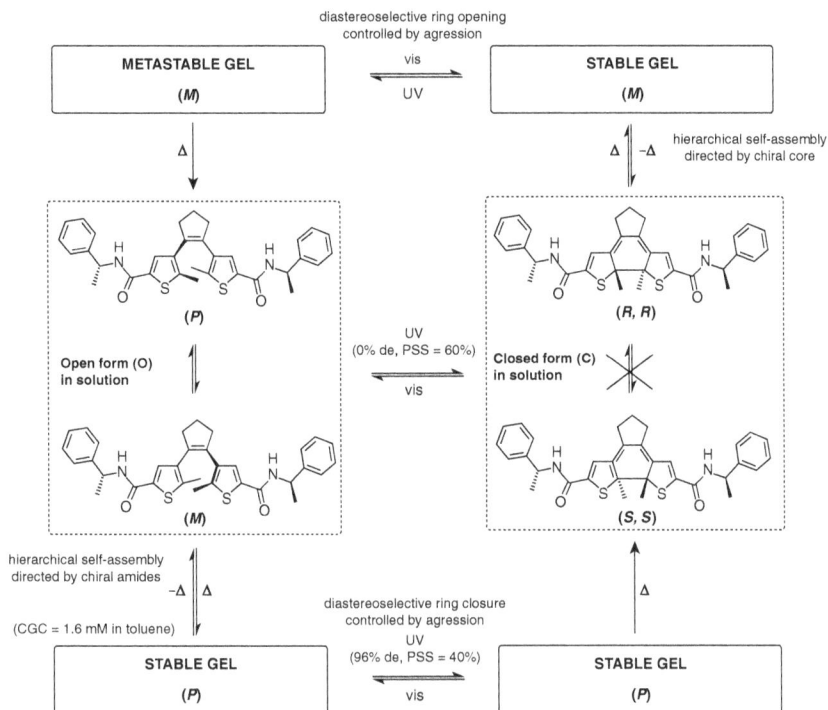

Figure 10.   The four-state cycle of a chiroptical switch. Combination of diastereoselective ring opening/closure processes with hierarchical self-assembly to allow reversible transcription of supramolecular into molecular chirality.[87] CGC = critical gelation concentration, de = diastereomeric excess, PSS = photostationary state, UV = ultraviolet light (313 nm), vis = visible light (>460 nm).

dependent on the size of the C_3-core and the mode of linkage between the wedged substituents and the core. This effect is seen, for example, in a series of hydrophobic C_3-symmetrical, discotic liquid crystalline materials comprising 2,2′-bipyridine-decorated benzene-1,3,5-tricarboxamide (BTC, Figure 11) central cores.[90,91] CD experiments performed by mixing achiral and chiral derivatives in hexane displayed strong amplification of optical activity. The quadruplet splitting reflection observed from the X-ray diffraction (XRD) measurements for both chiral and achiral materials was ascribed to the helical pitch associated with the chiral columns, confirming the intrinsic supramolecular order present in these C_3-symmetrical disc-shaped compounds. Evidence from UV–Vis absorption and CD spectroscopies and XRD measurements revealed that

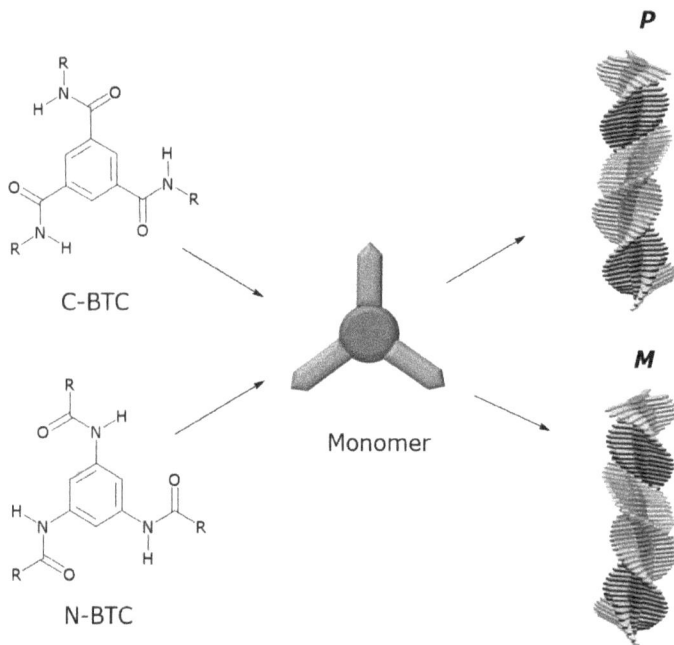

Figure 11.   Two constitutional isomers for BTCs that are the monomers for helical discotic supramolecular polymers.

chiral mesophase with columnar structure were preserved in non-polar solvents such as hexane, while no CD effect was observed in chloroform, where the molecule is in the molecularly dissolved state. This design was further extended by decorating BTC derivatives with, among other things, naphthalene diimides (NDI)[92] and porphyrins[93,94] possessing increased aromatic surface areas. The large π-conjugated surface drives the formation of intermolecular self-assembly in solution exploiting strong $\pi-\pi$ interactions and solvophobic effects.

Irrespective of the existence of a stereogenic center in these molecules, an important prerequisite for chiral amplification in supramolecular aggregates is the presence of an intrinsic conformational chirality in the assembly.[19] This conformation resembles a propeller, which is a chiral object. When the individual propellers are stacked on top of each other, the relative twist in one with respect to the next will favor the homochiral packing arrangement, resulting in a helical conformation. The three-fold intermolecular hydrogen-bonding between the amide groups of the central BTC core can drive the formation of helix. Even if equal amounts of *P* and *M* helices are present in the helical stacks of achiral derivatives, the addition of small amounts of chiral "sergeants" dictate the helical conformation of achiral "soldiers". This process of fast addition of a small amount of chiral compound to the achiral one is followed by a subsequent amplification in the CD effect as a result of chirality transfer, suggestive of highly dynamic supramolecular self-assembly resulting from the rapid exchange between the molecules in different stacks in solution. In the growth regime, the isodesmic or cooperative nature of the assembly is strictly dependent on the composition and constitution of the C_3 molecules.[95] Computational studies indicate that the dipole moment existing in the BTC aggregates is partially responsible for the cooperative nature of the assembly.[96]

The importance of the complementary secondary interactions in driving the formation of well-defined chiral self-assembled aggregates was demonstrated by examining the aggregation behavior of structurally analogous C_3 symmetrical molecules. Theoretical investigations on N-centered (N-BTC) and –C=O-centered (C-BTC) molecules (Figure 11) revealed that compared to the Ph−C=O bond, the rotation around Ph−N-H bond requires higher energy.[97] Hence, N-BTCs display less cooperation in self-assembly relative to C-BTCs explaining a lower degree of amplification of chirality in the former than the latter. Variations of the molecular structure, such as the distance of the stereogenic center to the central cores, the number of chiral side chains, as well as the central core, have been performed to rationalize the chirality transfer mechanism in C3-symmetrical self-assembled aggregates. A negligible energy difference between the diastereomerically correlated right- and left-handed helical aggregates was observed by replacing a hydrogen atom by deuterium.[98]

Substituting the peripheral alkoxy substituents with hydrophilic oligo-(ethylene oxide) side chains resulted in a two-step aggregation process upon heating in a polar solvent such as *n*-butanol, in contrast to the chiral self-assembly in hexane.[99] Results from neutron scattering, CD, and fluorescence measurements revealed the existence of two different kinds of supramolecular assemblies, (i) one which expresses chirality at the supramolecular level and (ii) one that does not, indicating that *n*-butanol is capable of interfering with the secondary interactions between the molecules and accounts for the positional order during the course of self-assembly.

The amide groups can be farther from the core and keep the self-assembly, and π-stacking of oligo(phenyleneethynylene) (OPE) C_3

compounds also show fascinating behavior. To examine the effect of the peripheral side chains linked to the aromatic core on chiroptical properties and supramolecular chirality, OPEs with various sidechains were studied.[100] The OPE-based tris-amides with a variable number of peripheral chiral side chains self-assembled into helical aggregates, but compounds without the amide group did not form a helical organization, proving the importance of this hydrogen bond in the ordering and transfer of chirality. The volume occupied by the sidechains in the form of chiral substituents has been studied in these compounds.[101] The number of stereogenic substituents in each molecule influences the helix reversal penalty, which increases with the number of chiral substituents, while the mismatch penalty (where an opposite handed monomer is incorporated in a majority stack) follows the opposite tendency. That tendency is opposite to BTAs. Therefore, the core structure is key, along with the sidechain chirality, in programming the chirality of the supramolecular polymers that are formed.

Molecules related to these but with lower symmetry form remarkable hierarchical helical structures, like the bis-amide in Figure 12.[102] Comprehensive microscopy and CD studies showed that the (*S*) enantiomer of the compound formed *M* helical nanofibers, while that in concentrated solutions coiled around one another to give *P* twisted superhelices. The fitting of the spectroscopic data showed that the assembly of the superstructures could not be explained by a cooperative model but rather was shown to behave in a way involving nucleation followed by elongation. The unpicking of complicated assembly pathways is vital for the understanding of this kind of hierarchical process, a feature that is also revealed in other systems.[103-106]

Apart from dense coiled structures, appropriate amphiphile design can lead to the formation of chiral nanotubes.[107] Often, these remarkable structures are the next hierarchical stage from helical intermediates.[108]

Achiral molecules have demonstrated the ability to form chiral nanotubes, a good example being that of an amphiphilic and zwitterionic cyanine dye (3,3′-bis(3-carboxy-*n*-propyl)-3,3′-di-*n*-octyl-5,5′,6,6′-tetrachlorobenzimida-carbocyanine, a.k.a. C8O3) that assembles when its ethanolic solution is added to an aqueous solution of sodium hydroxide.[109] The dye molecules are arranged in a twisted herringbone-type structure that gives rise to characteristic splitting in its spectra. The structure of the aggregates was found to be a bilayer tubule according to the TEM and diffraction experiments, and the nature of the dye's aggregate morphology was

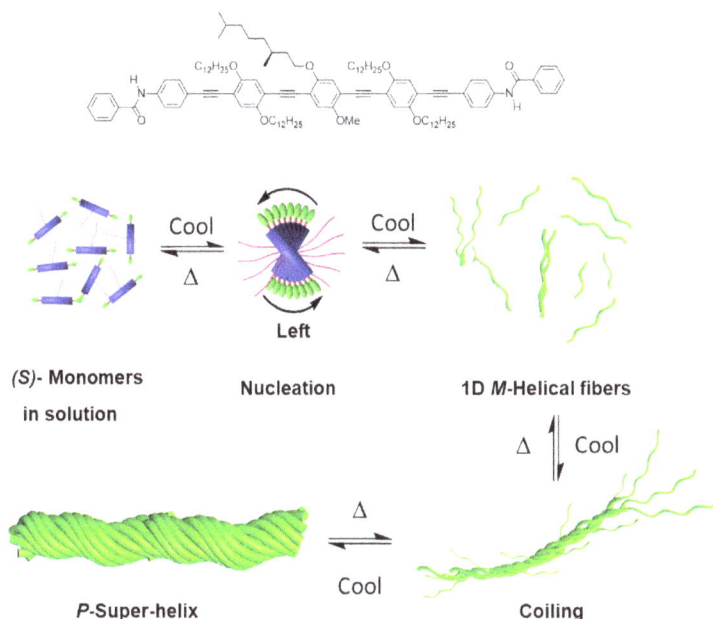

Figure 12.   Mechanism of self-assembly of a super-helix by the molecule shown where opposite helices are formed at subsequent hierarchical levels (with thanks to Vakayil K. Praveen and Ayyappanpillai Ajayaghosh for the cartoons used here).

dependent on the length of the hydrophobic chain attached to the core of the molecule.[110] The optical activity of these nanotube aggregates was dissected using Mueller polarimetry that separates the various contributions to optical activity.[111] Monitoring the light's polarization systematic during growth allowed detection of stages in the self-assembly. The hydrophobic effect was shown to have a determining effect in the mirror symmetry breaking taking place in the most initial steps of the aggregation pathway.

Chiral nanotubes are typically formed by surfactants, and the composition and constitution of these systems can determine what kind of structure results, a good example being that of peptide amphiphiles where the nature of the amino acid residues and the hydrophobic chain can be varied at will.[112] Here, flat ribbons or twisted structures are determined by the number of valine–glutamic acid dimeric units at the hydrophilic head of the amphiphile, and pH can also be used to influence the dimensions of the objects. Indeed, using pH to control hierarchical transfer of chirality is an interesting prospect.

A particularly unusual example of nanotube formation is that seen for a macrocycle comprising a hexa-4-phenylene rod and a chiral poly(ethylene oxide) chain.[113] This compound first assembles into ribbons and then curls around to give tubes that show uniform 20 nm diameter and 4.7 nm pitch. The nanophase separation of flexible and rigid polar and π-rich areas was believed to be the driving force for the formation of the tubes.

Perhaps the most remarkable self-assembled nanotubes to date are those formed by gemini structures incorporating a hexa-perihexabenzo-coronene core with chains of opposite polarity on two sides of this aromatic node.[114] These tubes can also be made chiral using the sergeants and soldiers approach, and the resulting objects are sufficiently robust that they can be oxidized to generate electrically conducting tubes.[115] Therefore, incorporation of aromatic functional units such as these promises to bring applications of these and related tubular soft matter structures.[116]

Synthetic covalent polymers have played an important role in stereochemistry, and many of the concepts in chirality transfer were actually developed in these macromolecules.[117] In particular the "sergeants and soldiers"[118] and "majority rules"[119] ideas were observed in poly(isocyanate) s. The transfer of chirality from chiral polymers to materials made from them has also been observed.[120,121] A particularly nice example is that of a block copolymer comprising a rigid poly(l-lactide) (PLLA) segment and an achiral and flexible polystyrene (PS) portion.[122] This material forms spiral superstructures hundreds of nanometers in diameter as a result of twisting between lamellae of the blocks. The understanding of the emergence of chiral morphologies in the block copolymer materials was probed spectroscopically, VCD showed the transfer of chirality from monomers to the intrachain conformation as well as to the interchain interactions.[123] There is a great deal that can be done and much to be studied in this kind of materials where understanding is relatively sparse compared with small molecule systems.

3.2. *Chiral monolayers and films*

The formation of chiral monolayers and films can be dominated by the interactions at interfaces and is dependent on the orientation of the molecules in this region and the strength and directionality of the interaction between the molecules and substrate.[124–127] Just as in three dimensions, racemates in 2D systems can form conglomerates or racemic compounds, although when interactions between substrate and adsorbate are strong the conglomerate is far more likely.[128]

Layers of lipids and related chiral surfactants on the surface of water, where amphiphiles are often oriented pseudo-perpendicular to the interface of the liquid with water, can lead to the emergence of macroscopically asymmetric objects.[129] The general area has been well reviewed for glycerol derived compounds.[130] Phospholipids were shown to generate morphologically chiral crystal-like objects in films.[131] Subsequently, other amphiphiles with hydrophilic head groups of differing polarity have shown similar phenomena and the non-covalent forces at work in bending the structures have been found to be principally electrostatic.[132] Thinking of how structural asymmetry propagates in solids, this information may prove important when applied to other systems. And it may not be necessary to make derivatives to find those structures that transfer chirality most, calculations were employed to predict the handedness of the domain shapes for these amphiphiles.[133]

Langmuir–Blodgett films of a helicenebisquinone beautifully show the effect of chirality and its transfer at the supramolecular level on an important physical property, in that case non-linear optical behavior.[134] The molecule self-assembles into fibers in its non-racemic form, where the helical compound packs into hexagonal columnar stacks,[135] and this material has huge optical activity.[136] The racemate does not form fibers under similar conditions. The chiral films comprising as many as sixty layers had a second harmonic light signal 1000 times the intensity of the racemic versions. It is believed that the chiral supramolecular organization — in addition to the intrinsic molecular chirality and polar order — is responsible for the extraordinarily good performance of the material. AFM showed that the non-racemic fibers were extremely homogeneous, an indication that the high order between fibers in this material is beneficial for the optical effect.

Achiral compounds can give rise to chiral morphologies at the air–water interface: spiral structures form on the surface of water in a Langmuir trough with an aliphatic barbituric acid derivative located there. Spirals of both handedness are observed, giving opposite CD effects on the transferred films.[137] The observation of hydrogen bonds by IR spectroscopy and the presence of H-aggregates indicated the presence of a bent-core supramolecular structure in the monolayers. Temperature and surface pressure affect the chirality as the film is formed, and the presence of achiral amphiphiles can stop the curling and therefore the development of the chiral structures.[138] While the emergence of chiral structures in pure systems is often remarkable, this sensitivity to additives and the robustness of the structural effects may be important when thinking of applications.

The transfer of Irving Langmuir films onto substrates can be done using the techniques developed with his collaborators Katharine B. Blodgett and Vincent J. Schaefer. The Langmuir–Blodgett requires compression of the film parallel to the interface, which could cause changes in the structure of fragile chiral structures, which is why the Langmuir–Schaefer method — that consists of a horizontal lift-off of the film at any stage of compression — is often the preferred method for chiral films. For example, the isomers of an achiral azobenzene can be transferred in its *cis* or *trans* form onto a solid substrate using the Langmuir–Schaefer method.[139] The films demonstrate high optical activity in the case of the *trans* isomer (not in the *cis* case), attributed to the ability of the molecule to adopt a helical arrangement. The same technique was used for the transfer of chiral poly(phenylacetylene) polymers onto graphite,[140] where the dynamic nature of the backbone was reflected in the diastereoisomeric domains observed on the substrate using AFM (Figure 13). These conditions for the growth of films are attractive for the hierarchical growth and transfer of chirality in a controlled way on a substrate, an interesting prospect for devices. The deposition conditions determine the quality and type of domain observed in these kinds of systems.[141]

Chirality can be transferred into the solutions that follows Langmuir monolayers when they template crystal growth. The oriented growth of glycine crystals can be templated by monolayers formed of amphiphiles derived from α amino acids so that the stereochemistry of the system is determined by the interface.[142] Opposite enantiotopic faces of the achiral glycine are favored by the surfactant enantiomers. Glutamic acid is stereoselectively incorporated onto this crystal (Figure 14).[143] Therefore, chirality transfer through oriented growth of an achiral compound can lead to hierarchical transfer of chirality at interfaces.

Many examples exist of the formation of chiral monolayers at interfaces, both from vacuum onto metal surfaces[124] and from solution.[126] The use of STM proved revolutionary in being able to image monolayers in systems as diverse as tartaric acid on copper[144] and liquid crystal molecules on graphite.[145] Both of these kinds of system can display transfer of chirality similar to solution systems, but where imaging allows observation of extremely fine detail in the structures.

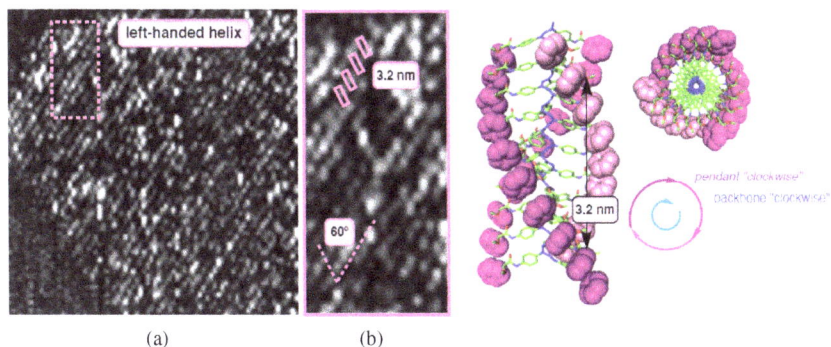

Figure 13. AFM images (a and b) of a chiral poly(phenylacetylene) transferred onto a graphite surface by the Langmuir–Schaefer technique, and a model of the helical structure that is observed (with thanks to Félix Freire for the graphic).

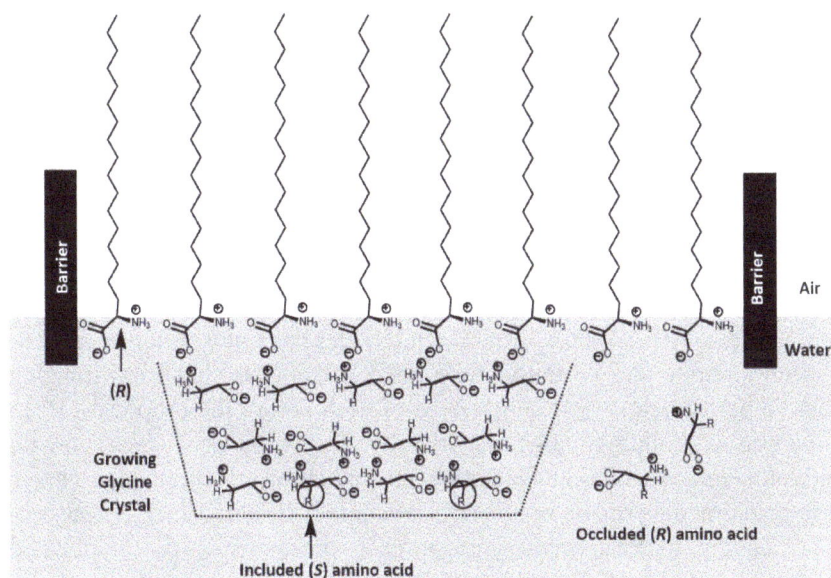

Figure 14. The templating of a glycine crystal formation by a chiral Langmuir layer and transfer of chirality to the enantioselective adsorption of an included amino acid.

The assembly of tartrate on copper is extremely sensitive to the optical purity of the adsorbate, and a drastic non-linear effect is observed whereby very small excesses of one enantiomer make the vast majority of the ordered domains the chirality of that isomer.[146] An entropic effect means that the minority enantiomer exists as a disordered material in between the well-organized domains of the majority domain. This observation is unique to STM, because observing poorly organized or amorphous material in the presence of organized chiral molecules is tremendously challenging, in solution for example. Another interesting case where chirality is not expressed over large dimensions is where enantiomers can inhibit the growth of each other's domains,[147] again only observable because STM "sees" every molecule on the surface. This situation could mean that poorly organized or minority species in solution, which are not visible in spectroscopic measurements, are actually playing a big role in the overall observed expressions of chirality.[148]

Relatively simple aromatics also show interesting hierarchical effects on metals in vacuum, for example 1-nitronaphthalene that chiral decamer aggregates and can actually be separated into their enantiomers with the STM tip,[149] or rubrene that has a chiral conformation that propagates through chains on the surface of gold,[150] and the assembly actually gives a complete monolayer with chiral domains.[151] Helicenes — compounds that are renowned for conglomerate formation in three dimensions in bulk crystals — also show interesting hierarchical stereochemistry phenomena on surfaces.[152–154] In particular, the remarkable observation that a monolayer of pentahelicene exists as a racemate but then upon nucleation of the second layer deracemize to give a conglomerate is intriguing.[155] It demonstrates beautifully how the growth of a surface can give rise to interesting rearrangements in the early stages of hierarchical growth.

Chiral systems at solid–solution interfaces have most often been studied on graphite and in non-volatile polar solvents so that STM imaging can be performed over a long period of time before the drop dries.[126] In common with other systems, remarkable effects have been observed with sub/molecular resolution, including the induction of chiral monolayers using achiral adsorbates and a chiral solvent,[156] diastereoselective adsorption leading to Pasteurian segregation,[157] mechanistically complicated sergeants and soldiers self-assembly effects (involving competitive adsorption, structural polymorphism, and adaptive host–guest recognition),[158] and the amplification of chirality in self-assembled bilayers.[159] The visualization of multiple layers with the STM can only go so far, as the technique relies on the tunneling of an electron between the probe tip and the conducting surface, but as we have discussed, it can provide unprecedented structural insight into processes taking place with sub/molecular resolution.

Following the hierarchical processes up into thin films is a tough challenge, and one where modern analytical techniques and emerging theory are bound to play a vital role. The characterization of the chiral structure in thin films can be complicated by the many optical effects that can give rise to apparent CD signals, although massive strides have been made in the characterization of films using Mueller Matrix analysis,[160] and there are many opportunities, as we shall discuss in a later section. The technique has proved useful for films.[161] The reason for interest in this area is that packing of charge carrying chromophores comprising the active layer of organic electronic devices can impact the efficiency of these devices significantly.[162,163] As CD does not require long-range order, it is particularly suited to the short-range order seen in these types of thin films. Briefly, it is worth noting that chiroptical properties often depend on film thickness,[164] many times because of the possibility of different organization and orientation at interfaces to the bulk, of which more will be discussed as follows.

3.3. *Chiral crystals*

Crystals can reflect one of the widest hierarchies of chiral transfer for synthetic materials, and chiral morphologies may be more common than one might think.[165] Studies on them are only now having something of a renaissance. Chiral crystals were documented in 1929, when it was claimed that 25% of simple molecular crystals have the ability to grow into helical structures.[166] Since then, a wide variety of materials from inorganic[167,168] to polymeric[169,170] to simple organic molecules[171,172] have been shown to form crystals with a twisted morphology. Of the 137 substances that formed twisted crystals documented in 1929, 135 were simple molecular substances, the vast majority of which grew in the form of banded spherulites.[173] When viewed between crossed polarizers in an optical microscope, banded spherulites show concentric rings with alternating extinction effects. It is thought that this phenomenon is evidence of crystals growing radially from a central nucleation point while twisting around their growth axis to produce helices. The color change seen in polarized optical micrographs is linked to the orientation of the lamellae and the respective extinction of that orientation (Figure 15). One of the most extensively studied polymeric materials that readily forms banded spherulites is polyethylene.[174–177]

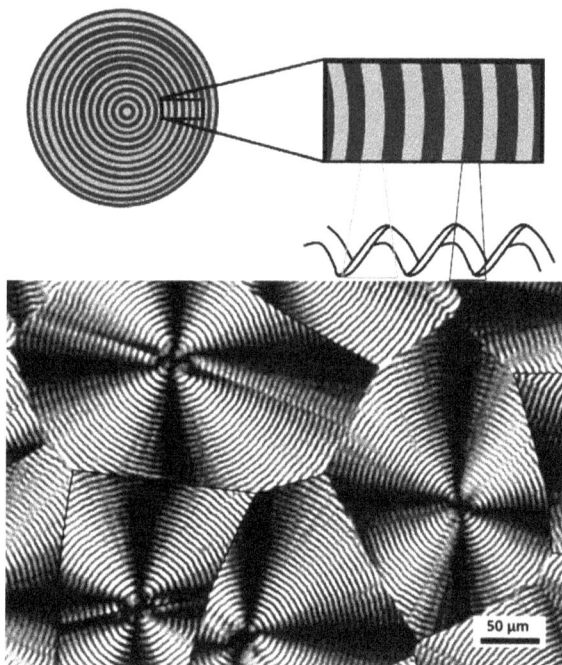

Figure 15.   Typical appearance of banded spherulites of polyethylene under a polarizing microscope. Reprinted with permission from the Ref. [177]. The American Chemical Society.

Characterizing the fine structure of these banded spherulites uses polarized optical microscopy, TEM, XRD, and scanning probe microscopy. There is a validated school of thought[178] believing that when band spacing is tight (<10 μm) the twisting is caused by chain tilts, resulting in congested packing and therefore generating large amounts of surface stress in lamellae that is relieved by twisting. This twisting can result in isochiral screw dislocations, which in turn can cause a positive feedback system which contributes to the twisting.

Many of the small molecule twisted crystals recorded are remarkably simple, such as urea, quinone, phenol, hippuric acid, and chlorobenzene.[166] Aspirin is one of the substances forming banded spherulites although impurities of salicylic acid (a precursor) are required in the melt to attain banded spherulites.[179] SEM images revealed that the spherulites were comprised of lamellar crystallites that were twisted along the growth

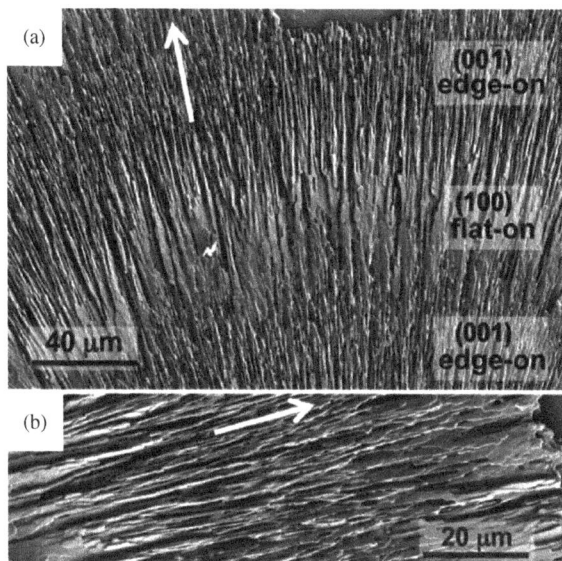

Figure 16. SEM images of aspirin spherulites (salicylic acid concentration in the melt 19 mol %; T = 20°C) showing overall arrangement of lamellae in the spherulite (a) and details of twisting (b). Growth direction is shown by arrows. Reprinted with permission from Ref. [179]. The American Chemical Society.

axis. (Figure 16) The stress associated with mixed crystal growth is believed to cause a twist moment in the growth of these twisted lamellae, and calculations of the strain caused by the presence of varying degrees of salicylic acid resulted in findings that were consistent with the experimental outcomes.

While twisted crystals are most commonly seen as banded spherulites, bent and twisted single crystals are surprisingly common. Naturally occurring minerals including quartz, rutile, stibnite, and pyrite have been observed in twisted forms.[165] However, the incredibly slow growth of minerals offer little opportunity for observation, and so there have been many attempts to mimic the crystallization of bent and twisted single crystals in the lab of both inorganic and organic materials (Figure 17). Zinc oxide nanoribbons,[180] potassium permanganate,[181] and lead sulfide[182] are examples of the former, while mannitol[183] and hippuric acid[173] are examples of organic materials that can form both banded spherulites and twisted single crystals. Benzamide, the first organic molecule for which

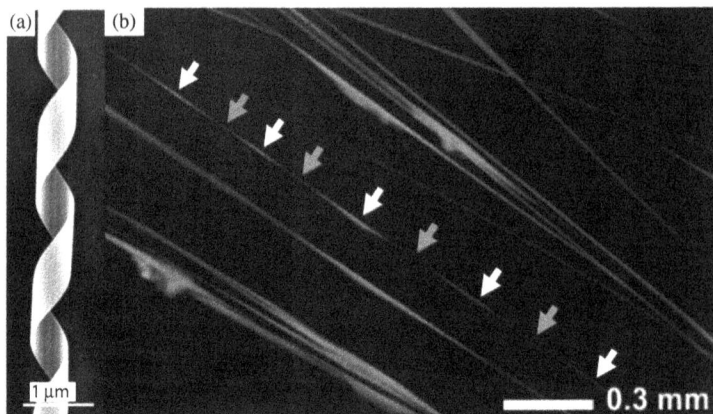

Figure 17.   (a) SEM image of superlattice-structured ZnO nanohelices. Reprinted with permission from the Ref. [180]. The American Chemical Society. (b) Isolated twisted needle of δ-mannitol grown from the melt. Edge-on and flat-on orientations are indicated by grey and white arrows, respectively. Reprinted with permission from the Ref. [183]. The American Chemical Society.

two polymorphs were discovered, has a twisted morphology in its meta-stable benzamide II form.[184] An intergrowth of lamellae with different crystallographic orientations causes stress that is likely to be the cause of the twisted morphology. Hierarchies of chiral twisting are apparent where molecular chirality is transferred to supramolecular chirality,[185,186] particularly in non-crystalline biological systems.[187,188] However, the link between supramolecular chirality and twisted crystals is where the chirality can be directly linked to the atomic configuration of the chiral molecules that form the twisted crystal.[165]

Decacyclene can form discrete crystals with a helical morphology, that both left- and right-handed objects were observed.[189] Single-crystal XRD showed that the solid has a chiral space group in which the compound has a gently twisted three-bladed propeller conformation. This distortion is from a planar structure by non-bonding interactions between hydrogen atoms at the peripheral naphthalene groups (Figure 18). It was proposed that left- and right-handed helical crystals correspond to opposite enantiomers of the molecular propeller, although proof is pending. Some link between stereochemistry and morphology is seen in thin films of poly-L-lactic acid (PLLA) and poly-D-lactic acid (PDLA): AFM

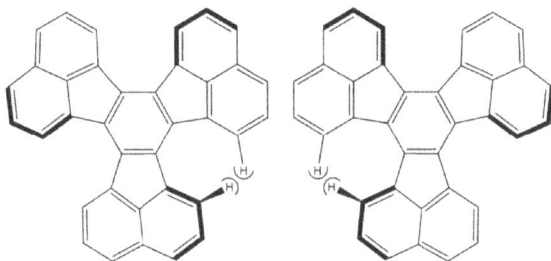

Figure 18. Conformational enantiomers of the molecular propeller of decacyclene.

showed that lamellae have enantiomeric morphology — S-shaped curvature for PLLA lamellae and Z-shaped curvature for PDLA.[190] The chiral polymer, poly(R-3-hydroxybutarate) forms banded spherulites where the lamellae always twists to the left.[191]

More usually, the connection between molecular chirality and aggregate asymmetry is enigmatic, and many mechanisms might be at work. Potassium dichromate forms both left- and right-handed twisted crystals when grown in a gel medium but when in the presence of L-or D- aspartic acid a preference for right- and left-handed crystals is shown, respectively.[3] Small changes in the molecular structures of some materials can cause a difference in handedness of the resulting crystals. In the case of (R)-(-)-4′-{w-[2-(4-hydroxy-2-nitrophenyloxy)-1-propyloxy]-1-alkyloxy}-4-biphenylcarboxylic acid, the number of methylene groups in the polymer chain backbone determines the handedness of the twisting.[192] When there are an odd number, the twisting is right-handed, while even numbers result in left-handed twisting.

So, if there seemingly no direct correlation between molecular chirality and the handedness of twisting in crystals, what is the cause? There are multiple current accepted mechanisms, each being applied for a variety of materials that form under their own certain set of conditions.[165] The one universal theme that unites these mechanisms is stress, from a variety of sources: (i) dislocations, including isochiral transverse screw dislocations, which, are sometime argued to be a form of stress relief rather than a cause and axial screw dislocations that can create elastic stress in the Eschelby mechanism, (ii) unbalanced surface stresses, particularly observed in polymers crystallized from the melt, where the polymer chains do not have time to reach an equilibrium configuration, resulting in surface stresses appearing, forcing lamellae to twist in order to relieve the unbalanced surface stress, and (iii) heterometry stresses, which are a product of sub-volumes of crystals growing at different times and there-fore having slightly different growth conditions, whether this is concen-tration of minor impurities, thermal gradients from inhomogeneous heat sources, or the latent heat of crystallization or density variations between the crystal and the melt from which it is grown. The result of these differ-ent growth conditions is that these sub-volumes can have slightly different compositions. Where two of these sub-volumes meet, there may be stress. It is believed that this stress, regardless of its source can be relieved by twisting or bending.

4. Influencing Hierarchical Chirality Transfer

The self-assembly of most supramolecular systems are sensitive to the exact conditions under which they are exposed, especially since these hierarchical processes are usually starting far from equilibrium.[193,194] Probably the most commonly exploited ways of promoting hierarchical assembly in synthetic systems are by simply changing temperature or solvent. The influence of vortexes and temperature gradients are also tremendously influential and will be discussed elsewhere in this vol-ume. Chemical reactivity is also a growing area, emulating biology, where assembly is initiated and/or propagated by covalent bond-making or -breaking. A nice example is the photo-initiated deprotection of a chlorophyll derivative that self-assembles under irradiation.[195] The length of the chiral nanotubular aggregates can be controlled with the number of photons, in a kind of non-equilibrium polymerization.[196] The combination of pushing systems away from equilibria by which-ever means and asymmetric self-assembly is at the heart of systems chemistry and its link with prebiotic chemistry.[197,198]

Many supramolecular systems can be extremely sensitive to changes in the solvent used to perform the self-assembly, because of the polarity of the solvent and corresponding change in solubility or because of the accessible temperature range with the solvent in question. The concentra-tion and rate of temperature change can also be important. For example, a non-amphiphilic tris(2,2′ bipyridinium)benzenetricarboxamide derivative bearing three tetrathiafulvalene (TTF) units each with two chains incorpo-rating point chirality assembles into columns in solution that show inver-sion of optical activity (Figure 19) when forming nanoscale fibers that subsequently coil into twisted microscopic fibers when allowed to cool quickly in a saturated solution in dioxane.[199]

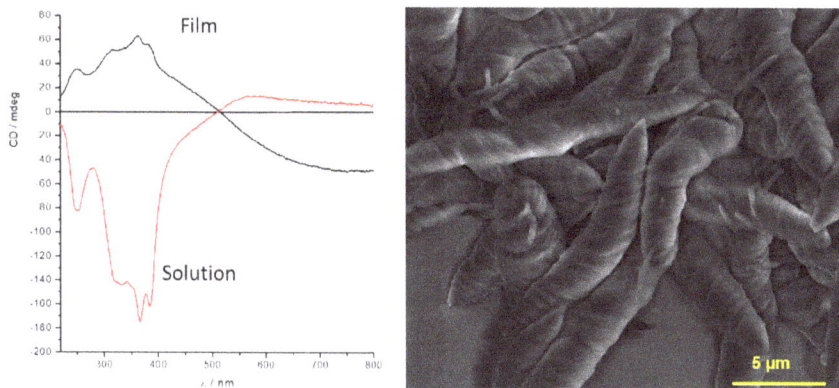

Figure 19.   CD spectra (left) of a film of a chiral C_3 TTF-containing compound on quartz deposited from dioxane solution at 10^{-4} M (positive bands at 350 nm) and a solution of the same compound in the same solvent at 5×10^{-5} M (negative band at 350 nm). The scanning electron micrograph on the right shows "microcroissants" formed by precipitation of the compound from dioxane when the saturated solution is cooled slowly.

When the saturated solution was cooled slowly, rather than fibers, finite microscale croissant-like objects were formed (Figure 19).[200] The twist sense of both the microscopic fibers and croissant-like objects is determined by the stereogenic center attached to the TTF residues, and the observation of nanofibers in both samples may suggest that the larger objects result from their coiling. At the microscopic level, molecular dynamics simulations indicated that single stacks would combine to generate a three-stranded coil and that the point chirality was transferred up to this level through the conformation of the bipyridine units to the helical columns. How this three-stranded system leads to the microscopic twisted objects and why dioxane is so specific in their formation await a full explanation. Interestingly, the effect of speed of gelation in a π-functional amphiphile was also observed, and remarkably with little solvent dependence.[201]

A solvent effect that leads to nanostructures with massively different optical activity was observed when a 1,1'-binaphthyl bis(perylenediimide) was precipitated from methylcyclohexane or chloroform: The fibers formed from the former solvent show a dissymmetry factor twice that of the spherical aggregates from the latter.[202] The CD and CPL of these aggregates varied with concentration and temperature in solution, shifting to longer wavelength and the CPL increasing with aggregation.

The extremely high luminescence dissymmetry was assigned to excitonic couplings between the perylenediimide units in the aggregates. A similar effect was seen in the hierarchical self-assembly of phthalhydrazide-bearing helicenes that form hydrogen-bonded trimers which subsequently stack into chiral fibers.[203] In that case, the luminescence dissymmetry ratio was higher for aggregates formed in chloroform over those formed in methanol. These works demonstrate the importance that morphology can have on a chiral property of the materials that self-assemble hierarchically, and the critical role that solvent choice has.

Circularly polarized light can be used to induce a specific asymmetric organization in materials (in addition to other phenomena discussed elsewhere in this volume). Molecular rotors based on overcrowded alkenes can be controlled in their helicity with circularly polarized light, and these chiral conformations can be transferred up to the twisted phases exhibited by the liquid crystal they are dissolved in.[204] That system is dynamic, but in a system where gold salts react to form a gold colloid, circularly polarized light irradiation during the reaction gives an assembly of colloids that is chiral and whose handedness is determined by the rotation of the light.[205] The use of asymmetric influences in hierarchical processes determined kinetically is surely an area of great opportunity.

5. Hierarchical Chirality Transfer in Multi-Component Systems

There is a growing interest in multi-component self-assembling mixtures and their development in systems chemistry,[206,207] where chemical tasks are carried out, although little has been done in the chiral arena, offering many opportunities to the many known multi-component chiral systems, some of which we discuss now. Multi-component supramolecular gel systems, in which more than one compound is added into the solvent, have been studied recently as they open up more prospects to produce new materials by modifying the properties of the single component gel. The self-assembly of two LMWG can lead to (i) self-sorting, (ii) random mixing, or (iii) specific co-assembly (Figure 20), which is very similar to the outcomes of gel formation of a racemic compound. One of the most explored areas of multi-component supramolecular gels is the mixing of two components that do not normally gel independently but form gels when combined.

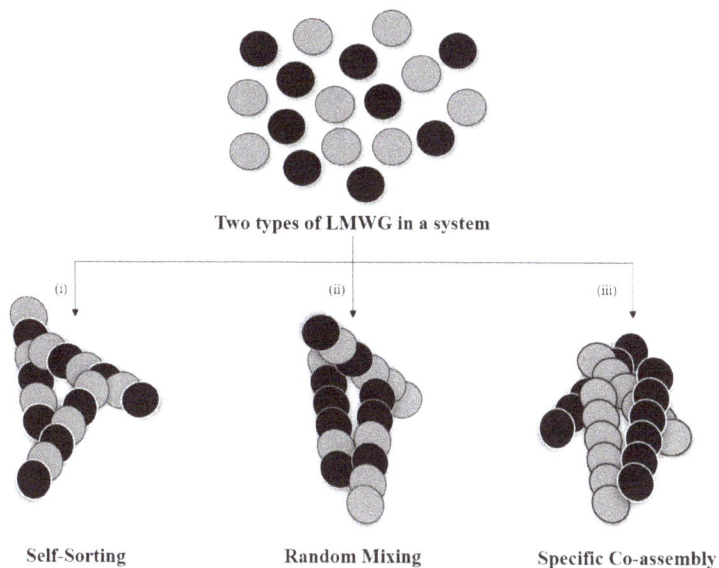

Figure 20. The possible outcomes of assembly of two different LMWG in cartoon form.

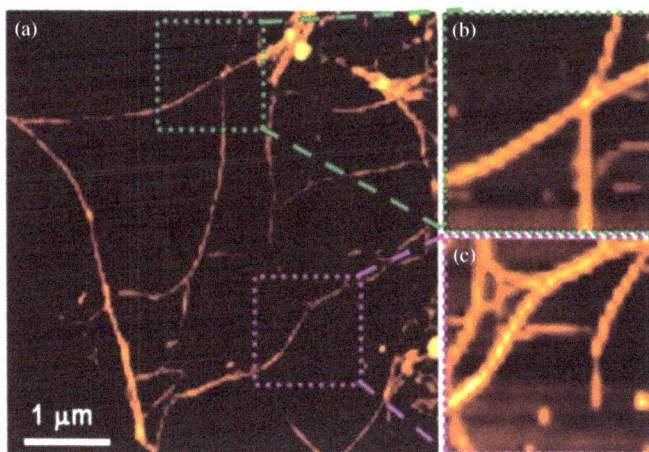

Figure 21. AFM micrographs of (a) racemic mixture of Fmoc-(L+D)Glu + (L+D)Lys, showing left- and right-handed helical fibers (b) zoomed version of right-handed helical fibers (a), and (c) zoomed version of left-handed helical fibers. Reprinted with permission from the Ref. [208].

Helical nanofibers can be formed by co-assembling two oppositely charged amino acid derivatives.[208] Fmoc-(L/D)Glu alone is unable to form a gel in water, and phosphate buffers or using the pH switch approach is needed (dissolving in basic pH and decreasing the pH using AcOH or GdL). However, the addition of (L)Lys to Fmoc-(L)Glu in water formed a supramolecular hydrogel upon heating and cooling *via* electrostatic attractions. AFM measurements revealed that equimolar gels of Fmoc-(D) Glu + (D)Lys and Fmoc-(L)Glu + (L)Lys show uniform right-handed and left-handed helical fibers, respectively. A four-component gel [Fmoc-(L+D)Glu + (L+D)Lys] showed the presence of both left- and right-handed helical fibers independently (Figure 21). It was observed that the components self-sort independently from each other and that the configuration of the stereogenic centers of the amino acids is a major factor for the chirality at the supramolecular level. In addition, the rheological properties of the tested multi-component gels shows that Fmoc-(D)Glu + (D) Lys and Fmoc-(L)Glu + (L)Lys have a higher storage modulus than the racemic mixture Fmoc-(L+D)Glu + (L+D)Lys.

A study of six chiral functionalized dipeptides, where four form self-supported gels as single component system, revealed a variety of behaviors dependent on the peptides.[209] They observed that mixing specific combinations showed (i) self-sorted gels with additive mechanical properties, (ii) self-sorted gels enabling late onset gelation, (iii) a disruptive self-sorted system, and (iv) a co-assembled gel with enhanced mechanical properties. In another family of compounds, the combination of mono- and di-benzylidene sorbitol gelators can immobilize the full range of water–ethanol mixtures (from 100% ethanol content to 100% water content) while neither gelator individually can.[210] The synergic effect was a result of mixing of the two molecules in lamellae in fibers that do not obviously show chiral morphology despite the enantiomeric purity of the natural starting material. The rheological data for the multi-component gels show higher values of modulus than the individual gelators and therefore indicate that the strength of the gel increases when the two gelators are incorporated in the same gel system.

Complementarity hydrogen bond donors and acceptors in separate components can be used to combine achiral and chiral compounds to transfer chirality. For example, combining a chiral dialkyl melamine derivative and a perylene bisimide leads to chiral stacks of dyes with hydrogen bonds directing the growth of the helical superstructure[211] (Figure 22). The dye's optical activity was revealed in induced CD in

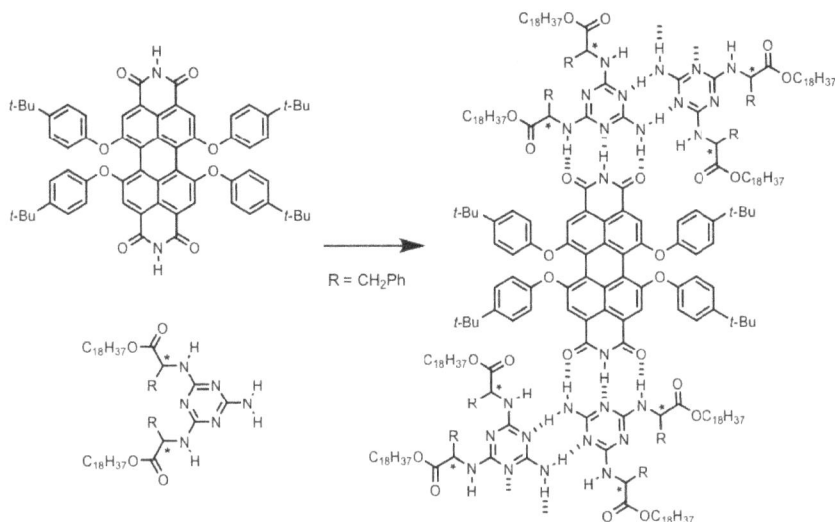

Figure 22. The supramolecular polymer formed by a chiral melamine derivative and a perylene diimide.

methylcyclohexane. Mesoscopic fibers form upon evaporation of the solvent. The strong imide–amine bonds favor formation of the mixed structure, as each component on their own can form aggregates through complementary hydrogen bonds. The stacking of the dye is favored even at very low concentrations, and the CD signals indicate their helicity.

The hierarchical growth of C_3-symmetrical superstructures form two-component achiral synthons comprising complementary hydrogen-bonding pairs — a melamine core (M) and three photoresponsive azobenzene fragments (Az-H and Az-OH) (Figure 23) — capable of engaging in non-covalent interactions between complexes M.(Az)$_3$ takes place in the initial formation of one-dimensional helices.[212] These strands bundle into higher order optically active supercoil structures, leading to spontaneous chiral amplification. The resulting long and helical fibers have intrinsic conformational chirality. The co-assembly in this two-component system is largely dictated by the mutually complementary hydrogen-bonding between carboxyl azobenzene derivatives and the melamine synthons. The photoisomerization of the azobenzene components in the self-assembled nanostructures imparts photoresponsiveness to the morphology and chirality of the supramolecular assemblies. The aromatic π-surfaces in the assembly tend to overlap through π–π stacking and, hence, the relative

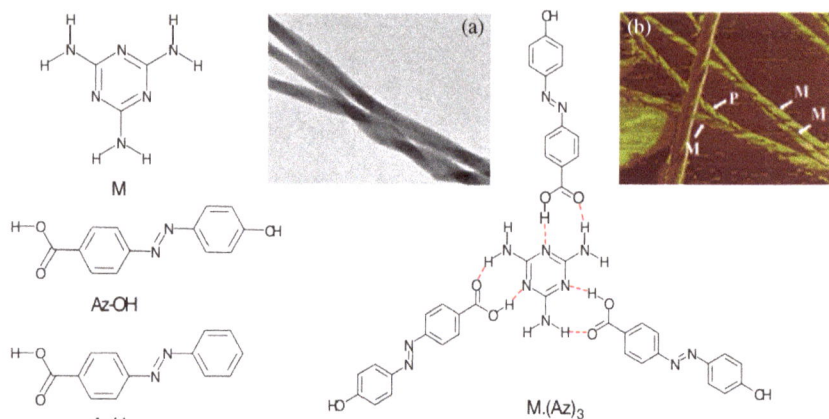

Figure 23.   The supramolecular polymer formed by: (a) chiral melamine derivative and (b) perylene diimide.

displacement of the aromatic rings within the molecular packing display a slight angle between the neighboring units, trigger a chiral bias, which is propagated further forming a helix in a certain direction, resulting in emergence and amplification of supramolecular chirality.

Rosette structures comprising complementary hydrogen-bonding groups lend themselves very well to the formation of chiral nanotubes, and this work has been reviewed very nicely.[213] Relatively simple surfactant-based nanotubes also show some intriguing effects. For example, a mixture of amphiphiles with different-length alkyl chains and opposite chirality polar head-groups assemble to give tubes whose chirality is dominated by the amphiphile with the shorter alkyl chains.[214] It is perhaps related with a "sergeants and corporals" effect seen in monolayers,[215] where a sergeant is ordered by a subordinate!

A very interesting series of lipids containing two polar groups — a chiral amino acid derivative and the other achiral amide — with a methylene chain between them form a series of self-assembled nanotubes whose size depends on the bulk of the amino acid residue.[216] These tubes were used as hosts for the preparation of a poly(diacetylene). The morphology of the polymer was determined by the nature of the nanotube, be it nanoparticles, nanotapes, nanocoils, or twisted nanofibers. It is a beautiful example of how a chiral container can template and control morphology of another material thanks to the hierarchical organization.

Charged surfactant self-assembling systems lend themselves very well to the preparation of hierarchical chirality transfer through the bending of the amphiphile arrays with chiral counterions.[217,218] The helical ribbons can act as templates for the formation of silica, providing hybrid materials[219] and also helical and twisted silica.[220] The structural chirality within the morphological chirality of these largely non-crystalline materials has been proven by VCD.[221] Very interestingly, the chirality that began in the organic template to silica formation can be passed into an achiral polyoxometallate.[222] This tremendous body of work demonstrates how a hierarchical process can be achieved in a stepwise manner, It is, perhaps, interesting to reflect upon how materials of this ilk could be prepared.

Organic fibers of a chiral nature can also template the hierarchical growth of colloidal structures that adopt a helical geometry, as shown elegantly for a phenylenediacetylene derivative capped with moieties containing point chiral polar groups at each end that template the formation a helical arrangement of gold nanoparticles.[223] Because the template is thermo-responsive, heating allows it to be removed, leaving the chiral gold colloid that shows a chiral plasmon band. This elegant approach demonstrates beautifully the hierarchical assembly by growth of colloids on colloids and might be used as an inspiration for the construction of other more complex materials using hierarchical strategies.

6. Opportunities

The growing community of scientists exploring hierarchically constructed chiral systems has several possible applications for the remarkable systems they can prepare. The area of sensing is a potentially important one for chiral structures[224–227] and may find use in systems where high sensitivity is required if correctly coupled to electronic or fluorescent outputs, where the chiral nature of the materials may provide an advantage over other systems. Recent proof of the ability of supramolecular assemblies to sense the chirality of amino acids[228] augurs well for the use of these unusual assemblies as sensors, and the development of robust structures constructed hierarchically offers many avenues available for exploration.

The chiral environment and unique surroundings of chemical functionalities in many of the materials makes them potentially useful as catalyst. A promising recent example showed how the BTC compounds can exploit the sergeants and soldiers principle to incorporate catalytically active copper centers on an achiral assembling unit that is able to perform asymmetric catalysis thanks to the environment provided by the twisted fiber.[229] Importantly, the loading with the chiral "sergeant" could be very low and favored the copper precatalyst's function in a hydrosilylation reaction. The ability to tune the surroundings of a catalyst in a supramolecular, reversible, and dynamic way could be an extremely useful approach for asymmetric catalysis of this type, where products of high enantiomeric purity were obtained.

During the development of hierarchical systems, beyond the initial stages of assembly the medium is not completely homogeneous, and therefore studies on the effects of assembly in low dimensions and under confinement become particularly relevant. The use of confinement in hierarchical assembly can play a very influential role in the chiral outcome, as shown in the mixing of a chiral ammonium salt and an anionic porphyrin under microfluidic conditions where very controlled diffusion conditions are obtained.[230] The chirality of the aggregates formed in a bulk phase under chaotic mixing is reversed over time, but with mixing under laminar flow conditions the initial chirality is maintained, even though the initial contact in the microfluidic chip is only for a fraction of a second. The work demonstrates how controlled flow can determine the outcome of a hierarchical assembly event from the initial nucleation events. Hierarchical processes to prepare larger-scale materials can take long periods of time, but this can be overcome in the case of cellulose nanocrystal organization using capillary confinement.[231] Chiral growth is controlled in this case by confined anisotropic evaporation of the suspension up a slide that leads to highly oriented material over many length scales and giving the dry film superior optical properties to a cast and dried film. Therefore, controlled evaporation can be a powerful tool in organizing chiral materials and allowing ordered hierarchical films.

Characterization of films can be extremely challenging but is important for understanding the properties of the materials. Habitual bench-top CD spectrometers have shown that the chirality and symmetry of organic molecules has an important effect on the efficiency of organic solar cells derived from these materials,[232] but spatial resolution is relatively poor. Synchrotron radiation CD with highly collimated light allows much more precise mapping of thin films.[233] Detection of multiple aggregation states in thin films of chiral 1,4-phenylene-based oligothiophenes in different proportions depending on the sample preparation is possible.[234] This work revealed the parameters that can be altered to affect the supramolecular structures within the thin films, opening the opportunity to study these materials and others with previously unknown precision.

7. Outlook

The transfer of chirality in materials is broadly applicable but, to date, has been explored scarcely, partially because of the difficulty in predicting the phenomena that take place across the hierarchy of scales. It is likely that the combined use of modeling and a variety of microscopies will aid in the understanding of hierarchical processes that take place over different timescales. Just as TEM has been used for the observation of twisted ribbons into helical ribbons,[235] so SEM can be used to observe higher levels of assembly.[236] It is clear that the complexity of greater hierarchical levels will require inventive ways to treat samples and to interpret the asymmetrical structures that are generated.

A challenge in the area of defining chirality of hierarchical morphologies lies in the identification of chiral phases using microscopy, since the optical activity (determined by relative orientation of chromophores) does not necessarily reflect the shape of the aggregates at the microscopic and macroscopic scale. For example, defining the chirality of a curved wing from a bird or airplane might be clear at a macroscopic scale with respect to the left or right side of animal or vehicle, but at a microscopic scale this kind of feature is challenging to define, particularly in materials without some degree of longer range order.

In terms of potential applications of hierarchical chirality transfer, it is perhaps interesting to be inspired by artificial mechanical metamaterials.[237] These structures can be used, for example, for impact energy absorption.[238] Indeed, there are mechanical properties of manufactured materials that are inaccessible through other designs, as is the case for three-dimensional mechanical metamaterials.[239] In natural biological systems, chirality comes inbuilt, perhaps in synthetic materials the kind of molecular design that has been applied to mechanical materials could be interesting. Inspiration from biological systems is inferred frequently[240,241] and the understanding of natural hierarchical structure pathways for chiral transfer is likely to give synthetic supramolecular scientists clues as to how to control this kind of process in human-made materials.

To reach these goals, fundamental insights still need to be made, for example in understanding the roles of solvents, which can be a powerful ally when used well as shown in the steric demand control of the assembly of metallo-supramolecular assemblies.[242] Another challenge in the area of chiral (photo)conducting materials is to control the organization and hierarchical assembly of the compounds so that the twist between chromophores is compatible with charge transport. Hydrogen bonds have been used successfully to assemble chiral aggregates of diketopyrrolopyrroles[243] that exhibit photoconductivity, although the incorporation of stereogenic centers in these systems and others does not aid the functional properties, at least under the conditions that were used.[244] It was pointed out astutely that the nature of the stereogenic center could be very influential in the order and therefore property, and the number of different surroundings to stereogenic centers employed widely is presently relatively limited given the scope for preparing chiral compounds. Alternatively, the use of chiral additives to influence hierarchical organization of functional materials[245] is an extremely promising effect that will benefit from deeper understanding.

There is no doubt that the preparation of composites where chirality is transferred between different kinds of nanoscale materials is a growing and important area that reminds us of transfer of asymmetry in biological systems. Carbon nanodots — carbon based materials that are soluble in water and are around 3 nm in diameter — can be prepared enantioselectively using readily available chiral starting materials and can transfer their chirality to assemblies of porphyrins.[246] The chirality of the dots is seen beautifully using VCD, and the transfer from the molecular to nanoscale by CD. The scope of these kinds of carbon materials is massive because of their potential compatibility with many materials, depending on the material used in the hydrothermal synthesis of the nanodots.

These, and other, applications[247,248] await for hierarchically controlled next-generation materials that are bound to provide improved and totally new functions.

Acknowledgments

The authors would like to thank all of the people that collaborate in their own research, some of which is mentioned in this review, and to the University of Nottingham. ARM thanks the funding received from the European Union's Horizon 2020 research and innovation program under the Marie Sklodowska-Curie grant agreement No:793424. We thank warmly Félix Freire, Vakayil K. Praveen, and Ayyapanpillai Ajayaghosh for kindly providing graphic material.

References

1. G. H. Wagnière, *On Chirality and the Universal Asymmetry: Reflections on Image and Mirror Image*, Wiley-VCH, Weinheim, Germany, (2007).
2. D. K. Kondepudi and D. J. Durand, Chiral asymmetry in spiral galaxies? *Chirality* **13**(7), 351–356 (2001).
3. H. Imai and Y. Oakia, Emergence of helical morphologies with crystals: Twisted growth under diffusion-limited conditions and chirality control with molecular recognition, *CrystEngComm* **12**(6), 1679–1687 (2010).
4. U. Hananela, A. Ben-Moshea, H. Diamanta, and G. Markovich, Spontaneous and directed symmetry breaking in the formation of chiral nanocrystals, *PNAS* **116**(23), 11159–11164 (2019).
5. V. F. Korolovych, V. Cherpak, D. Nepal, A. Ng, N. R. Shaikh, A. Grant, R. Xiong, T. J. Bunning, and V. V. Tsukruk, Cellulose nanocrystals with different morphologies and chiral properties, *Polymer* **145**(1), 334e347 (2018).
6. A. Narkevicius, L. M. Steiner, R. M. Parker, Y. Ogawa, B. Frka-Petesic, and S. Vignolini, Controlling the self-assembly behavior of aqueous chitin nanocrystal suspensions, *Biomacromolecules* **20**(7), 2830–2838 (2019).
7. D. B. Amabilino, ed., *Chirality at the Nanoscale*, Wiley-VCH, Weinheim, Germany, (2009).
8. M. Goh, S. Matsushita, and K. Akagi, From helical polyacetylene to helical graphite: Synthesis in the chiral nematic liquid crystal field and morphology-retaining carbonization, *Chem. Soc. Rev.* **39**(7), 2466–2476 (2010).
9. F. R. Keene, ed., *Chirality in Supramolecular Assembles: Causes and Consequences*, Wiley, Chichester, UK, (2017).
10. F. Pop, N. Zigon, and N. Avarvari, Main-Group-Based electro- and photoactive chiral materials, *Chem. Rev.* **119**(14), 8435–8478 (2019).
11. B. A. Grzybowski, C. E. Wilmer, J. Kim, K. P. Browne, and K. J. M. Bishop, Self-assembly: From crystals to cells, *Soft Matter.* **5**(6), 1110–1128 (2009).
12. M. A. Mateos-Timoneda, M. Crego-Calama, and D. N. Reinhoudt, Supramolecular chirality of self-assembled systems in solution, *Chem. Soc. Rev.* **33**(6), 363–372 (2004).
13. M. Liu, L. Zhang, and T. Wang, Supramolecular chirality in self-assembled systems, *Chem. Rev.* **115**(15), 7304–7397 (2015).
14. E, Yashima, N. Ousaka, D. Taura, K. Shimomura, T. Ikai, and K. Maeda, Supramolecular helical systems: Helical assemblies of small molecules,

foldamers, and polymers with chiral amplification and their functions, *Chem. Rev.* **116**(22), 13752–13990 (2016).

15. B. Yue, L. Yin, W. Zhao, X. Jia, M. Zhu, B. Wu, S. Wu, and L. Zhu, Chirality transfer in coassembled organogels enabling wide-range naked-eye enantiodifferentiation, *ACS Nano* **13**(11), 12438–12444 (2019).

16. J. M. Ribo and D. Hochberg, Spontaneous mirror symmetry breaking: An entropy production survey of the racemate instability and the emergence of stable scalemic stationary states, *Phys. Chem. Chem. Phys.* **22**(25), 14013–14025 (2020).

17. L. Pérez-García and D. B. Amabilino, Spontaneous resolution under supra-molecular Control, *Chem. Soc. Rev.* **31**(6), 342–356 (2002).

18. L. Pérez-García and D. B. Amabilino, Spontaneous resolution, whence and whither: From enantiomorphic solids to chiral liquid crystals, monolayers and macro- and supra-molecular polymers and assemblies, *Chem. Soc. Rev.* **36**(6), 941–967 (2007).

19. E. W. Meijer and A. R. A. Palmans, Amplification of chirality in dynamic supramolecular aggregates, *Angew. Chem. Int. Ed.* **46**(47), 8948–8968 (2007).

20. S. M. Morrow, A. J. Bissette, and S. P. Fletcher, Transmission of chirality through space and across length scales, *Nat. Nanotechnol.* **12**(5) 410–419 (2017).

21. L. J. Prins, P. Timmerman, and D. N. Reinhoudt, Amplification of chirality: The "Sergeants and Soldiers" principle applied to dynamic hydrogen-bonded assemblies, *J. Am. Chem. Soc.* **123**(42) 10153–10163 (2001).

22. R. D. Ennulat and A. J. Brown, Mesomorphism of homologous series 2. Odd–even effect, *Mol. Cryst. Liq. Cryst.* **12**(4) 367–378 (1971).

23. S. Bualek, S. Patumtevapibal, and J. Siripitayananon, Odd–even effect in the helical twisting power of chiral molecules in induced cholesteric meso-phases, *Chem. Phys. Lett.* **79**(2), 389–391 (1981).

24. R. J. Miller and H. F. Gleeson, The influence of pretransitional phenomena on blue phase range, *Liq. Cryst.* **14**(6), 2001–2011 (1993).

25. A. T. M. Marcelis, A. Koudijs, and E. J. R. Sudhölter, Odd–even effects in the optical properties of chiral twin liquid-crystalline cholesteryl ω-(cyanobi-phenylyloxy) alkanoates, *Rec. Trav. Chim. Pays-Bas* **113**(11), 524–526 (1994).

26. D. B. Amabilino, J.-L. Serrano, T. Sierra, and J. Veciana, Long-range effects of chirality in aromatic poly(isocyanide)s, *J. Polym. Sci., Polym. Chem.* **44**(10), 3161–3174(2006).

27. J. W. Goodby, E. Chin, T. M. Leslie, J. M. Geary, and J. S. Patel, Helical twist sense and spontaneous polarization direction in ferroelectric smectic liquid crystals, *J. Am. Chem. Soc.* **108**(16), 4729–4735 (1986).

28. E. Ramos-Lermo, B. M. W. Langeveld-Voss, R. A. J. Janssen, and E. W. Meijer, Odd–even effect in optically active poly(3,4-dialkoxythiophene, *Chem. Commun.* **35**(9), 791–792 (1999).

29. S. S. Panda, K. Shmilovich, A. L. Ferguson, and J. D. Tovar, Controlling Supramolecular chirality in peptide–π–peptide networks by variation of the alkyl spacer length, *Langmuir* **35**(43), 14060–14073 (2019).
30. T. Sanji, N. Kato, M. Kato, and M. Tanaka, Helical folding in a helical channel: Chiroptical transcription of helical information through chiral wrapping, *Angew. Chem. Int. Ed.* **44**(44), 7301–7304 (2005).
31. N. Numata and S. Shinkai, 'Supramolecular wrapping chemistry' by helix-forming polysaccharides: A powerful strategy for generating diverse polymeric nano-architectures, *Chem. Commun.* **47**(7), 1961–1975 (2011).
32. D. J. Hill, M. J. Mio, R. B. Prince, T. S. Hughes, and J. S. Moore, A field guide to foldamers, *Chem. Rev.* **101**(12), 3893–4012 (2001).
33. H. Juwarker, J.-m. Suk, and K.-S. Jeong, Foldamers with helical cavities for binding complementary guests, *Chem. Soc. Rev.* **38**(12), 3316–3325 (2009).
34. R. B. Prince, S. A. Barnes, and J. S. Moore, Foldamer-Based molecular recognition, *J. Am. Chem. Soc.* **122**(12), 2758–2762 (2000).
35. P. Mateus, N. Chandramouli, C. D. Mackereth, B. Kauffmann, Y. Ferrand, and I. Huc, Allosteric recognition of homomeric and heteromeric pairs of monosaccharides by a foldamer capsule, *Angew. Chem. Int. Ed.* **59**(14), 5797–5805 (2020).
36. Q. Gan, Y. Wang, and H. Jiang, Twisted helical microfibers by hierarchical self-assembly of an aromatic oligoamide foldamer, *Chin. J. Chem.* **31**(5), 651–656 (2013).
37. W. Cai, G.-T. Wang, P. Du, R.-X. Wang, X.-K. Jiang, and Z.-T. Li, Foldamer organogels: A circular dichroism study of glucose-mediated dynamic helicity induction and amplification, *J. Am. Chem. Soc.* **130**(40), 13450–13459 (2008).
38. V. F. Korolovych, V. Cherpak, D. Nepal, A. Ng, N. R. Shaikh, A. Grant, R. Xiong, T. J. Bunning, and V. V. Tsukruk, Cellulose nanocrystals with different morphologies and chiral properties, *Polymer* **145**, 334–347 (2018).
39. R. M. Parker, G. Guidetti, C. A. Williams, T. Zhao, A. Narkevicius, S. Vignolini, and B. Frka-Petesic, The self-assembly of cellulose nanocrystals: Hierarchical design of visual appearance, *Adv. Mater.* **30**(19), 1704477 (2018).
40. L. Pasteur, Recherches sur les relations qui peuvent exister entre la forme cristalline, la composition chimique et le sens de la polarisation rotatoire, *Ann. Chim. Phys.* **24**, 442–459 (1848).
41. R. G. Kostyanovsky, Louis Pasteur did it for us especially, *Mendeleev Commun.* **13**(3), 85–90 (2003).
42. Y. Tobe, The reexamination of Pasteur's experiment in Japan, *Mendeleev Commun.* **13**(3), 93–94 (2003).
43. N. Berova, L. Di Bari, and G. Pestcitelli, Application of electronic circular dichroism in configurational and conformational analysis of organic compounds, *Chem. Soc. Rev.* **36**(6), 914–931 (2007).

44. G. Pestcitelli, L. Di Bari, and N. Berova, Application of electronic circular dichroism in the study of supramolecular systems, *Chem. Soc. Rev.* **43**(15), 5211–5233 (2014).
45. H. Hashizume, M. Shiraki, and K. Imahori, Study of circular dichroism of proteins and polypeptides in relation to their backbone and side chain conformations, *J. Biochem.* **62**(5), 543–551 (1967).
46. Y. Wei, A. A. Thyparambil, and R. A. Latour, Protein helical structure determination using CD spectroscopy for solutions with strong background absorbance from 190 to 230 nm, *Biochim. Biophys. Acta — Proteins and Proteomics* **1844**(12), 2311–2337 (2014).
47. J. K. Maurus and G. R. Bird, Circular dichroism of sensitizing dye aggregates, *J. Phys. Chem.* **76**(21), 2982–2986 (1972).
48. G. Scheibe, O. Wörz, F. Haimerl, W. Seiffert, and J. Winkler, Circular dichroism spectra of adsorbed dye-aggregates, *J. Chim. Phys.* **65**(1), 146–151 (1968).
49. L. Stryer and E. R. Blout, Optical rotatory dispersion of dyes bound to macromolecules. Cationic dyes: Polyglutamic acid complexes, *J. Am. Chem. Soc.* **83**(6), 1411–1418 (1961).
50. R. Kuroda and T. Honma, CD spectra of solid-state samples, *Chirality* **12**(4), 269–277 (2000).
51. M. Minguet, D. B. Amabilino, K. Wurst, and J. Veciana, Circular dichroism studies of crystalline chiral and achiral α-nitronyl nitroxide radicals in KBr matrix, *J. Chem. Soc. Perkin Trans.* **2**, 670–676 (2001).
52. F. Formaggio, M. Crisma, C. Toniolo, and J. Kamphuis, Solid-state CD and peptide helical screw sense, *Biopolymers* **38**(3), 301–304 (1996).
53. G. Albano, G. Pescitelli, and L. Di Bari, Chiroptical properties in thin films of π-conjugated systems, *Chem. Rev.* **120**(18), 10145–10243 (2020).
54. J. Liu, H. Su, L. Meng, Y. Zhao, C. Deng, J. C. Y. Ng, P. Lu, M. Faisal, J. W. Y. Lam, X. Huang, H. Wu, K. S. Wong, and B. Z. Tang, What makes efficient circularly polarised luminescence in the condensed phase: Aggregation-induced circular dichroism and light emission, *Chem. Sci.* **3**(9), 2737–2747 (2012).
55. J. Roose, B. Z. Tang, and K. S. Wong, Circularly-Polarized luminescence (CPL) from chiral AIE molecules and macrostructures, *Small.* **12**(47), 6495–6512 (2016).
56. Y. T. Sang, J. L. Han, T. H. Zhao, P. F. Duan, and M. H. Liu, Circularly polarized luminescence in nanoassemblies: Generation, amplification, and application, *Adv. Mater.* **32**, 1900110 (2019).
57. P. J. Stephens, Theory of vibrational circular dichroism, *J. Phys. Chem.* **89**(5), 748–752 (1985).
58. J. Sadlej, J. Cz. Dobrowolski, and J. E. Rode, VCD spectroscopy as a novel probe for chirality transfer in molecular interactions, *Chem. Soc. Rev.* **39**(5), 1478–1488 (2010).

59. V. Setnička, M. Urbanová, K. Volka, S. Nampally, and J.-M. Lehn, Investigation of guanosine-quartet assemblies by vibrational and electronic circular dichroism spectroscopy, a novel approach for studying supramolecular entities, *Chem. Eur. J.* **12**(34), 8735–8743 (2006).

60. M. Linares, A. Minoia, P. Brocorens, D Beljonne, and R. Lazzaroni, Expression of chirality in molecular layers at surfaces: Insights from modelling, *Chem. Soc. Rev.* **38**(3), 806–816 (2009).

61. A. Kasperski and P. Szabelski, Theoretical modeling of surface confined chiral nanoporous networks: Cruciform molecules as versatile building blocks, *Chirality* **27**(7), 397–404 (2015).

62. A. Painelli, F. Terenziani, L. Angiolini, T. Benelli, and L. Giorgini, Chiral interactions in azobenzene dimers: A combined experimental and theoretical study, *Chem. Eur. J.* **11**(20), 6053–6063 (2005).

63. G. Pescitelli, L. Di Bari, Lorenzo, and N. Berova, Conformational aspects in the studies of organic compounds by electronic circular dichroism, *Chem. Soc. Rev.* **40**(9), 4603–4625 (2011).

64. S. Marchesan, L. Waddington, C. D. Easton, D. A. Winkler, L. Goodall, J. Forsyth, and P. G. Hartley, Unzipping the role of chirality in nanoscale self-assembly of tripeptide hydrogels, *Nanoscale* **4**(21), 6752–6760 (2012).

65. C. Oliveras-González, M. Linares, D. B. Amabilino, N. Avarvari, Large synthetic molecule that either folds or aggregates through weak supramolecular interactions determined by solvent, *ACS Omega* **4**(6), 10108–10120 (2019).

66. C. C. Lee, C. Grenier, E. W. Meijer, and A. P. H. J. Schenning, Preparation and characterization of helical self-assembled nanofibers, *Chem. Soc. Rev.* **38**(3), 671–683(2009).

67. A. W. Weitkamp, The acidic constituents of degras. A new method of structure elucidation, *J. Am. Chem. Soc.* **67**(3), 447–454 (1945).

68. T. Tachibana and H. Kambara, Enantiomorphism in the helical aggregate of lithium 12-hydroxystearate, *J. Am. Chem. Soc.* **87**(13), 3015–3016 (1965).

69. A. Brizard, R. Oda, and I. Huc, Chirality effects in self-assembled fibrillar networks, *Top. Curr. Chem.* **256**, 167–218 (2005).

70. P. Terech and R. G. Weiss, Low molecular mass gelators of organic liquids and the properties of their gels, *Chem. Rev.* **97**(8), 3133–3160 (1997).

71. P. Dastidar, Supramolecular gelling agents: Can they be designed? *Chem. Soc. Rev.* **37**(12), 2699–2715 (2008).

72. E. R. Draper and D. J. Adams, How should multi-component supramolecular gels be characterized, *Chem. Soc. Rev.* **47**(10), 3395–3405 (2018).

73. D. B. Amabilino, D. K. Smith, and J. W. Steed, Supramolecular materials, *Chem. Soc. Rev.* **46**(9), 2404–2420 (2017).

74. D. J. Abdallah and R. G. Weiss, n-Alkanes gel n-alkanes (and many other organic liquids), *Langmuir* **16**, 352–355 (2000).

75. A. Ajayaghosh, V. K. Praveen, and C. Vijayakumar, Organogels as scaffolds for excitation energy transfer and light harvesting, *Chem. Soc. Rev.* **37**(1), 109–122 (2008).

76. D. K. Smith, Lost in translation? Chirality effects in the self-assembly of nanostructured gel-phase materials, *Chem. Soc. Rev.* **38**(3), 684–694 (2009).

77. L. A. Estroff and A. D. Hamilton, Water gelation by small organic molecules, *Chem. Rev.* **104**(3), 1201–1218 (2004).

78. M. Jokic, J. Makarevic, and M. Zinic, A novel type of small organic gelators: Bis(amino acid) oxalyl amides, *J. Chem. Soc., Chem. Commun.* **31**(17), 1723–1724 (1995).

79. X. Luo, B. Liu, and Y. Liang, Self-assembled organogels formed by monochain L-alanine derivatives, *Chem. Commun.* (17), 1556–1557 (2001).

80. L. Frkanec and M. Žinić, Chiral bis(amino acid)- and bis(amino alcohol)-oxalamidegelators. Gelation properties, self-assembly motifs and chirality effects, *Chem. Commun.* **46**(4), 522–537 (2010).

81. U. Maitra, V. K. Potluri, N. M. Sangeetha, P. Babu, and A. R. Raju, Helical aggregates from a chiral organogelator, *Tetrahedron Asymmetry* **12**(3), 477–480 (2001).

82. Y. Kageyama, T. Ikegami, N. Hiramatsu, S. Takeda, and T. Sugawara, Structure and growth behavior of centimeter-sized helical oleate assemblies formed with assistance of medium-length carboxylic acids, *Soft Matter* **11**(18), 3550–3558 (2015).

83. L. L. E. Mears, E. R. Draper, A. M. Castilla, H. Su, Zhuola, B. Dietrich, M. C. Nolan, G. N. Smith, J. Doutch, S. Rogers, R. Akhtar, H. Cui, and D. J. Adams, Drying affects the fiber network in low molecular weight hydrogels, *Biomacromolecules* **18**(11), 3531–3540 (2017).

84. D. A. Tómasson, D. Ghosh, Z. Kržišnik, L. H. Fasolin, A. A. Vicente, A. D. Martin, P. Thordarson, and K. K. Damodaran, Enhanced mechanical and thermal strength in mixed-enantiomers-based supramolecular gel, *Langmuir* **34**(43), 12957–12967 (2008).

85. S. Hecht, Optical switching of hierarchical self-assembly: Towards "Enlightened" materials, *Small* **1**(1), 26–28 (2003).

86. J. Kim, J. W. Kim, H. C. Kim, L. Zhai, H.-U Ko, and R. M. Muthoka, Review of soft actuator materials, *Int. J. Precision Eng. Manufact.* **20**(12), 2221–2241 (2019).

87. J. J. D. de Jong, L. N. Lucas, R. M. Kellogg, J. H. van Esch, and B. L. Feringa, Reversible optical transcription of supramolecular chirality into molecular chirality, *Science* **304**(5668), 278–281 (2004).

88. S. E. Gibson and M. P. Castaldi, Applications of chiral C_3-symmetric molecules, *Chem. Commun.* (29), 3045–3062 (2006).

89. F. Wang, M. A. J. Gillissen, P. J. M. Stals, A. R. A. Palmans, and E. W. Meijer, Hydrogen bonding directed supramolecular polymerisation of

oligo(phenylene-ethynylene)s: Cooperative mechanism, core symmetry effect and chiral amplification, *Chem. Eur. J.* **18**(37), 11761–11770 (2012).

90. A. R. A. Palmans, J. A. J. M. Vekemans, E. E. Havinga, and E. W. Meijer, Sergeants-and-Soldiers principle in chiral columnar stacks of disc-shaped molecules with C_3 symmetry, *Angew. Chem. Int. Ed. Engl.* **36**(23), 2648–2651 (1997).

91. S. Cantekin, T. F. A. de Greef, and A. R. A. Palmans, Benzene-1,3,5-tricarboxamide: A versatile ordering moiety for supramolecular chemistry, *Chem. Soc. Rev.* **41**(18), 6125–6137 (2012).

92. B. Narayan, C. Kulkarni, and S. J. George, Synthesis and self-assembly of a C_3-symmetric benzene-1,3,5-tricarboxamide (BTA) anchored naphthalene diimide disc, *J. Mater. Chem. C* **1**(4), 626–629 (2013).

93. R. van Hameren, A. M. van Buul, M. A. Castriciano, V. Villari, N. Micali, P. Schön, S. Speller, L. Monsù-Scolaro, A. E. Rowan, J. A. A. W. Elemans, and R. J. M. Nolte, Supramolecular porphyrin polymers in solution and at the solid–liquid interface, *Nano Lett.* **8**(1), 253–259 (2008).

94. N. Veling, R. van Hameren, A. M. van Buul, A. E. Rowan, R. J. M. Nolte, and J. A. A. W. Elemans. Solvent-dependent amplification of chirality in assemblies of porphyrin trimers based on benzene tricarboxamide, *Chem. Commun.* **48**(36), 4371–4373 (2012).

95. C. Kulkarni, E. W. Meijer, and A. R. A. Palmans, Cooperativity scale: A structure–mechanism correlation in the self-assembly of benzene-1,3,5-tricarboxamides, *Acc. Chem. Res.* **50**(8), 1928–1936 (2017).

96. D. B. Korlepara, W. R. Henderson, R. K. Castellano, and S. Balasubramanian, Differentiating the mechanism of self-assembly in supramolecular polymers through computation, *Chem. Commun.* **55**(26), 3773–3776 (2019).

97. P. J. M. Stals, J. C. Everts, R. de Bruijn, I. A. W. Filot, M. M. J. Smulders, R. Martin-Rapun, E. A. Pidko, T. F. A. de Greef, A. R. A. Palmans, and E. W. Meijer, Dynamic supramolecular polymers based on benzene-1,3,5-tricarboxamides: The influence of amide connectivity on aggregate stability and amplification of chirality, *Chem. Eur. J.* **16**(3), 810–821 (2010).

98. S. Cantekin, D. Balkenende, M. Smulders, A. R. A. Palmans, and E. W. Meijer, The effect of isotopic substitution on the chirality of a self-assembled helix. *Nat. Chem.* **3**(11), 42–46 (2011).

99. L. Brunsveld, H. Zhang, M. Glasbeek, J. A. J. M. Vekemans, and E. W. Meijer, Hierarchical growth of chiral self-assembled structures in protic media, *J. Am. Chem. Soc.* **122**(26), 6175–6182 (2000).

100. F. García and L. Sánchez, Structural rules for the chiral supramolecular organization of OPE-based discotics: Induction of helicity and amplification of chirality, *J. Am. Chem. Soc.* **134**(1), 734–742 (2012).

101. E. E. Greciano, J. Calbo, J. Buendía, J. Cerdá, J. Aragó, E. Ortí, and L. Sánchez, Decoding the consequences of increasing the size of self-assembling

tricarboxamides on chiral amplification, *J. Am. Chem. Soc.* **141**(18), 7463–7472 (2019).

102. M. Hifsudheen, R. K. Mishra, B. Vedhanarayanan, V. K. Praveen, and A. Ajayaghosh, The helix to super-helix transition in the self-assembly of π-systems: Superseding of molecular chirality at hierarchical level, *Angew. Chem. Int. Ed. Engl.* **56**(41), 12634–12638 (2017).

103. M. F. J. Mabesoone, A. J. Markvoort, M. Banno, T. Yamaguchi, F. Helmich, Y. Naito, E. Yashima, A. R. A. Palmans, and E. W. Meijer, Competing interactions in hierarchical porphyrin self-assembly introduce robustness in pathway complexity, *J. Am. Chem. Soc.* **140**(25), 7810–7819 (2018).

104. E. E. Greciano, B. Matarranz, and L. Sánchez, Pathway complexity versus hierarchical self-assembly in N-annulated perylenes: Structural effects in seeded supramolecular polymerization, *Angew. Chem. Int. Ed. Engl.* **57**(17), 4697–4701 (2018).

105. I. Helmers, M. Niehues, K. K. Kartha, B. Jan Ravoo, and G. Fernández, Synergistic repulsive interactions trigger pathway complexity, *Chem. Commun.* **56**(63), 8944–8947 (2020).

106. J. Matern, K. K. Kartha, L. Sánchez, and G. Fernández, Consequences of hidden kinetic pathways on supramolecular polymerization, *Chem. Sci.* **11**(26), 6780–6788 (2020).

107. T. Shimizu, M. Masuda, and H. Minamikawa, Supramolecular nanotube architectures based on amphiphilic molecules, *Chem. Rev.* **105**(4), 1401–1443 (2005).

108. T. G. Barclay, K. Constantopoulos, and J. Matisons, Nanotubes self-assembled from amphiphilic molecules *via* helical intermediates, *Chem. Rev.* **114**(20), 10217–10291 (2014).

109. U. De Rossi, S. Dähne, S. C. J. Meskers, and H. P. J. M. Dekkers, Spontaneous formation of chirality in J-aggregates showing davydov splitting, *Angew. Chem. Int. Ed.* **35**(7), 760–763 (1996).

110. H. von Berlepsch, C. Böttcher, A. Ouart, C. Burger, S. Dähne, and S. Kirstein, Supramolecular structures of J-aggregates of carbocyanine dyes in solution, *J. Phys. Chem. B* **104**(22), 5255–5262 (2000).

111. A. Thomas, T. Chervy, S. Azzini, M. Li, J. George, C. Genet, and T. W. Ebbesen, Mueller polarimetry of chiral supramolecular assembly, *J. Phys. Chem. C* **122**(25), 14205–14212 (2018).

112. T. J. Moyer, H. Cui, and S. I. Stupp, Tuning nanostructure dimensions with supramolecular twisting, *J. Phys. Chem. B* **117**(16), 4604–4610 (2013).

113. W.-Y. Yang, E. Lee, and M. Lee, Tubular organization with a coiled ribbon from an amphiphilic rigid-flexible macrocycle, *J. Am. Chem. Soc.* **128**(11), 3484–3485 (2006).

114. W. Jin, T. Fukushima, M. Niki, A. Kosaka, N. Ishii, and T. Aida, Self-assembled graphitic nanotubes with one-handed helical arrays of a chiral amphiphilic molecular graphene, *PNAS* **102**(31), 10801–10806 (2005).

115. T. Yamamoto, T. Fukushima, A. Kosaka, W. Jin, Y. Yamamoto, N. Ishii, and T. Aida, Conductive one-handed nanocoils by coassembly of hexabenzocoronenes: Control of morphology and helical chirality, *Angew. Chem. Int. Ed.* **47**(9), 1672–1675 (2008).

116. T. Shimizu, W. Ding, and N. Kameta, Soft-Matter nanotubes: A platform for diverse functions and applications, *Chem. Rev.* **120**(4), 2347–2407 (2020).

117. M. M. Green, J.-W. Park, T. Sato, A. Teramoto, S. Lifson, R. L. B. Selinger, and J. V. Selinger, The macromolecular route to chiral amplification, *Angew. Chem. Int. Ed.* **38**(21), 3138–3154 (1999).

118. M. M. Green, M. P. Reidy, R. D. Johnson, G. Darling, D. J. O'Leary, and G. Willson, Macromolecular stereochemistry: The out-of-proportion influence of optically active comonomers on the conformational characteristics of polyisocyanates. The sergeants and soldiers experiment, *J. Am. Chem. Soc.* **111**(16), 6452–6454 (1989).

119. M. M. Green, B. A. Garetz, B. Munoz, H. P. Chang, S. Hoke, and R. G. Cooks, Majority rules in the copolymerization of mirror image isomers, *J. Am. Chem. Soc.* **117**(14) 4181–4182 (1995).

120. R. M. Ho, Y. W. Chiang, S. C. Lin, and C. K. Chen, Helical architectures from self-assembly of chiral polymers and block copolymers, *Prog. Polym. Sci.* **36**(3), 376–453 (2011).

121. Y.-W. Chiang, R.-M. Ho, C. Burger, and H. Hasegawa, Helical assemblies from chiral block copolymers, *Soft Matter* **7**(21), 9797–9803 (2011).

122. K.-C. Yang and R.-M. Ho, Spiral hierarchical superstructures from twisted ribbons of self-assembled chiral block copolymers, *ACS Macro Lett.* **9**(8), 1130–1134 (2020).

123. H.-F. Wang, K.-C. Yang, W.-C. Hsu, J.-Y. Lee, J.-T. Hsu, G. M. Grason, E. L. Thomas, J.-C. Tsai, and Rong-Ming Ho, Generalizing the effects of chirality on block copolymer assembly, *PNAS* **116**(10), 4080–4089 (2019).

124. S. M. Barlow and R. Raval, Complex organic molecules at metal surfaces: Bonding, organisation and chirality, *Surf. Sci. Rep.* **50**(6–8), 201–341 (2003).

125. R. Fasel, M. Parschau, and K. H. Ernst, Chirality transfer from single molecules into self-assembled monolayers, *Angew. Chem. Int. Ed.* **42**(42), 5178–5181 (2003).

126. J. A. A. W. Elemans, I. De Cat, H. Xu, and S. De Feyter, Two-dimensional chirality at liquid–solid interfaces, *Chem. Soc. Rev.* **38**(3), 722–736 (2009).

127. F. Zaera, Chirality in adsorption on solid surfaces, *Chem. Soc. Rev.* **46**(23), 7374–7398 (2017).

128. S. Dutta and A. J. Gellman, Enantiomer surface chemistry: Conglomerate versus racemate formation on surfaces, *Chem. Soc. Rev.* **46**(24), 7787–7839 (2017).

129. N. Nandi and D. Vollhardt, Effect of molecular chirality on the morphology of biomimetic Langmuir monolayers, *Chem, Rev.* **103**(10), 4033–4075 (2003).
130. R. J. Craven and R. W. Lencki, Polymorphism of acylglycerols: A stereochemical perspective, *Chem. Rev.* **113**(10) 7402–7420 (2013).
131. R. M. Weis and H. M. McConnell, Two-dimensional chiral crystals of phospholipid, *Nature* **310**, 47–49 (1984).
132. K. Thirumoorthy, N. Nandi, and D. Vollhardt, Role of electrostatic interactions for the domain shapes of Langmuir monolayers of monoglycerol amphiphiles, *J. Phys. Chem. B* **109**(21), 10820–10829 (2005).
133. K. Thirumoorthy, N. Nandi, and D. Vollhardt, Prediction of the handedness of the domains of monolayers of D-N-palmitoyl aspartic acid: Integrated molecular orbital and molecular mechanics based calculation, *Colloids Surf. A — Physicochem. Eng Asp.* **282**, 222–226 (2006).
134. T. Verbiest, S. Van Elshocht, M. Kauranen, L. Hellemans, J. Snauwaert, C. Nuckolls, T. J. Katz, and A. Persoons, Strong enhancement of nonlinear optical properties through supramolecular chirality, *Science* **282**(5390), 913–915 (1998).
135. A. J. Lovinger, C. Nuckolls, and T. J. Katz, Structure and morphology of helicene fibers, *J. Am. Chem. Soc.* **120**(2) 264–268 (1998).
136. C. Nuckolls, T. J. Katz, G. Katz, P. J. Collings, and L. Castellanos, Synthesis and aggregation of a conjugated helical molecule, *J. Am. Chem. Soc.* **121**(1), 79–88 (1999).
137. X. Huang, C. Li, S. Jiang, X. Wang, B. Zhang, and M. Liu, Self-Assembled spiral nanoarchitecture and supramolecular chirality in Langmuir–Blodgett films of an achiral amphiphilic barbituric acid, *J. Am. Chem. Soc.* **126**(5), 1322–1323 (2004).
138. X. Huang and M. Liu, Regulation of supramolecular chirality and morphology of the LB film of achiral barbituric acid by amphiphilic matrix molecules, *Langmuir* **22**(9), 4110–4115 (2006).
139. Y. Zhang, P. Chen, and M. Liu, Supramolecular chirality from an achiral azobenzene derivative through the interfacial assembly: Effect of the configuration of azobenzene unit, *Langmuir* **22**(24), 10246–10250 (2006).
140. R. Rodríguez, J. Ignés-Mullol, F. Sagués, E. Quiñoá, R. Riguera, and F. Freire, Helical sense selective domains and enantiomeric superhelices generated by Langmuir–Schaefer deposition of an axially racemic chiral helical polymer, *Nanoscale* **8**(6), 3362–3367 (2016).
141. F. Freire, E. Quiñoá, and R. Riguera, Chiral nanostructure in polymers under different deposition conditions observed using atomic force microscopy of monolayers: Poly(phenylacetylene)s as a case study, *Chem. Commun.* **53**(3), 481–492 (2017).
142. E. M. Landau, M. Levanon, L. Leiserowitz, M. Lahav, and J. Sagiv, Transfer of structural information from Langmuir monolayers to three-dimensional growing crystals, *Nature* **318**(6044), 353–356 (1985).

143. E. M. Landau, S. G. Wolf, M. Levanon, L. Leiserowitz, M. Lahav, and J. Sagiv, Stereochemical studies in crystal nucleation. Oriented crystal growth of glycine at interfaces covered with Langmuir and Langmuir-Blodgett films of resolved alpha.-amino acids, *J. Am. Chem. Soc.* **111**(4), 1436–1445 (1989).

144. M. O. Lorenzo, C. J. Baddeley, C. Muryn, and R. Raval, Extended surface chirality from supramolecular assemblies of adsorbed chiral molecules, *Nature* **404**(6776), 376–379 (2000).

145. D. M. Walba, F. Stevens, N. A. Clark, and D. C. Parks, Detecting molecular chirality by scanning tunneling microscopy, *Acc. Chem. Res.* **29**(12), 591–597 (1996).

146. S. Haq, N. Liu, V. Humblot, A. P. J. Jansen, and R. Raval, Drastic symmetry breaking in supramolecular organization of enantiomerically unbalanced monolayers at surfaces, *Nature Chem.* **1**(5), 409–414 (2009).

147. A. Robin, P. Iavicoli, K. Wurst, M. S. Dyer, S. Haq, D. B. Amabilino, and R. Raval, A racemic conglomerate nipped in the bud: A molecular view of enantiomer cross-inhibition of conglomerate nucleation at a surface, *Cryst. Growth Des.* **10**(10), 4516–4525 (2010).

148. R. Raval, Molecular assembly at surfaces: Progress and challenges, *Faraday Discuss.* **204**, 9–33 (2017).

149. M. Bohringer, K. Morgenstern, W. D. Schneider, and R. Berndt, Separation of a racemic mixture of two-dimensional molecular clusters by scanning tunneling microscopy, *Angew. Chem. Int. Ed.* **38**(6), 821–823 (1999).

150. M.-C. Blüm, E. Ćavar, M. Pivetta, F. Patthey, and W.-D. Schneider, Conservation of chirality in a hierarchical supramolecular self-assembled structure with pentagonal symmetry, *Angew. Chem. Int. Ed.* **44**(33), 5334–5337 (2005).

151. L. Wang, H. Kong, X. Song, X. Liub, and H. Wang, Chiral supramolecular self-assembly of rubrene, *Phys. Chem. Chem. Phys.* **12**(44), 14682–14685 (2010).

152. K.-H. Ernst, Stereochemical recognition of helicenes on metal surfaces, *Acc. Chem. Res.* **49**(6), 1182–1190 (2016).

153. J. D. Fuhr, M. W. van der Meijden, L. J. Cristina, L. M. Rodríguez, R. M. Kellogg, J. E. Gayone, H. Ascolani, and M. Lingenfelder, Chiral expression of adsorbed (MP) 5-amino[6]helicenes: From random structures to dense racemic crystals by surface alloying, *Chem. Commun.* **53**(1), 130–133 (2017).

154. J. Seibel, M. Parschau, and K.-H. Ernst, Double layer crystallization of heptahelicene on noble metal surfaces, *Chirality* **32**(7), 975–980 (2020).

155. A. Mairena, L. Zoppi, J. Seibel, A. F. Tröster, K. Grenader, M. Parschau, A. Terfort, and K. H. Ernst, Heterochiral to homochiral transition in pentahelicene 2D crystallization induced by second-layer nucleation, *ACS Nano* **11**(1), 865–871 (2017).

156. I. Destoop, A. Minoia, O. Ivasenko, A. Noguchi, K. Tahara, Y. Tobe, R. Lazzaroni, and S. De Feyter, Transfer of chiral information from a chiral solvent to a two-dimensional network, *Faraday Discuss.* **204**, 215–231 (2017).

157. H. Xu, W. J. Saletra, P. Iavicoli, B. Van Averbeke, E. Ghijsens, K. S. Mali, A. P. H. J. Schenning, D. Beljonne, R. Lazzaroni, D. B. Amabilino, and S. De Feyter, Pasteurian segregation on a surface imaged *in situ* at the molecular level, *Angew. Chem. Int. Ed.* **51**(48), 11981–11985 (2012).

158. H. Cao, I. Destoop, K. Tahara, Y. Tobe, K. S. Mali, and S. De Feyter, Complex chiral induction processes at the solution/solid interface, *J. Phys. Chem. C* **120**(31), 17444–17453 (2016).

159. H. Cao and S. De Feyter, Amplification of chirality in surface-confined supramolecular bilayers, *Nature Comm.* **9**, 3416 (2018).

160. J. H. Freudenthal, E. Hollis, and B. Kahr, Imaging chiroptical artifacts, *Chirality* **21**(1E), E20–E27 (2009).

161. J. Zablocki, O. Arteaga, F. Balzer, D. Hertel, J. J. Holstein, G. Clever, J. Anhäuser, R. Puttreddy, K. Rissanen, K. Meerholz, A. Lützen, and M. Schiek, Polymorphic chiral squarine crystallites in textured thin films, *Chirality* **32**(5), 619–631 (2020).

162. C. Wang, H. Dong, W. Hu, Y. Liu, and D. Zhu, Semiconducting π-conjugated systems in field-effect transistors: A material odyssey of organic electronics, *Chem. Rev.* **112**(4), 2208–2267, (2012).

163. P. M. Beaujuge and J. M. J. Fréchet, Molecular design and ordering effects in π-functional materials for transistor and solar cell applications, *J. Am. Chem. Soc.* **113**(50), 20009–20029 (2011).

164. M. R. Craig, P. Jonkheijm, S. C. J. Meskers, A. P. H. J. Schenning, and E. W. Meijer, The chiroptical properties of a thermally annealed film of chiral substituted polyfluorene depend on film thickness, *Adv. Mater.* **15**(17), 1435–1438 (2003).

165. A. G. Shtukenberg, Y. O. Punin, A. Gujral, and B. Kahr, Growth actuated bending and twisting of single crystals, *Angew. Chem. Int. Ed.* **53**(3), 672–699 (2014).

166. F. Bernauer, *"Gedrillte" Kristalle: Verbreitung, Entstehungsweise und Beziehungen zu optischer Aktivitat und Molekulasymmetrie*, Gebrüder Borntraeger, Berlin, (1929).

167. O. Giraldo, S. L. Brock, M. Marquez, S. L. Suib, H. Hillhouse, and M. Tsapatsis, Spontaneous formation of inorganic helices, *Nature* **405**(6782), 38–38 (2000).

168. M. Yang and N. Kotov, Nanoscale helices from inorganic materials, *J. Mater. Chem.* **21** (19), 6775–6792 (2011).

169. C. Li, D. Yan, S. Cheng, F. Bai, T. He, L. Chien, F. W. Harris, and B. Lotz, Double-Twisted helical lamellar, crystals in a synthetic main-chain chiral

polyester similar to biological polymers, *Macromolecules* **32**(2), 524–527 (1999).

170. C. Kübel, D. P. Lawrence, and D. C. Martin, Super-helically twisted strands of poly(m-phenylene isophthalamide) (MPDI), *Macromolecules* **34**(26), 9053–9058 (2001).

171. D. M. Ho and R. A. Pascal, Jr., Decacyclene: A molecular propeller with helical crystals, *Chem. Mater.* **5**(9), 1358–1361 (1993).

172. A. G. Shtukenberg, A. Gujral, E. Rosseeva, X. Cui, and B. Kahr, Mechanic of twisted hippuric acid crystals untwisting as they grow, *Cryst. Eng. Comm.* **17**(46), 8817–8824 (2015).

173. A. Shtukenberg, E. Gunn, M. Gazzano, J. Freudenthal, E. Camp, R. Sours, E. Rosseeva, and B. Kahr, Bernauer's bands, *ChemPhysChem* **12**(8), 1558–1571 (2011).

174. Y. Fujiwara, The superstructure of melt-crystallised polyethylene. I. Screwlike orientation of unit cell in polyethylene spherulites with periodic extinction rings, *J. Appl. Polym. Sci.* **4**(10), 10–15 (1960).

175. A. Keller, Investigations on banded spherulites, *J. Polym. Sci.* **39**(135), 151–173 (1959).

176. H. Keith and F. Padden, Banding in polyethylene and other spherulites, *Macromolecules* **29**(24), 7776–7786 (1996).

177. A. Lovinger, Twisted crystals and the origin of banding in spherulites of semicrystalline polymers, *Macromolecules* **53**(3), 741–745 (2020).

178. H. Keith and F. Padden, Banding in polyethylene and other spherulites, *Macromolecules* **29**(24), 7776–7786, (1996).

179. X. Cui, A. I. Rohl, A. Shtukenberg, and B. Kahr, Twisted aspirin crystals, *J. Am. Chem. Soc.* **135**(9), 3395–3398 (2013).

180. P. Gao, W. Mai, and Z. L. Wang, Superelasticity and nanofracture mechanics of ZnO nanohelices, *Nano Lett.* **6**(11), 2536–2543 (2006).

181. H. Imai and Y. Oaki, Emergence of morphological chirality from twinned crystals, *Angew. Chem. Int. Ed.* **43**(11), 1363–1368 (2004).

182. M. J. Bierman, Y. K. A. Lau, A. V. Kvit, A. L. Schmitt, and S. Jin, Dislocation-Driven nanowire growth and eshelby twist, *Science* **320**(5879), 1060–1063 (2008).

183. A. G. Shtukenberg, X. Cui, H. Freudenthal, E. Gunn, E. Camp, and B. Kahr, Twisted mannitol crystals establish, homologous growth mechanisms for high-polymer and small-molecular ring-banded spherulites, *J. Am. Chem. Soc.* **134**(14), 6354–6364 (2012).

184. A. G. Shtukenberg, R. Drori, E. Sturm, N. Vidavsky, A. Haddad, J. Zheng, L. Estroff, H. Weissman, S. G. Wolf, E. Shimoni, C. Li, N. Fellah, E. Efrati, and B. Kahr, Crystals of benzamide, the first polymorphous molecular compound, are helicoidal, *Angew. Chem. Int. Ed.* **59**(34), 14593–14601 (2020).

185. C. Y. Li, S. Z. D. Cheng, J. J. Ge, F. Bai, J. Z. Zhang, I. K. Mann, L. Chien, F. W. Harris, and B. Lotz, Molecular orientations in flat- elongated and helical lamellar crystals of a main-chain nonracemic chiral polyester, *J. Am. Chem. Soc.* **122**(1), 72–79 (2000).

186. Y. Yang, Y. Zhang, and Z. Wei, Supramolecular Helices: Chirality transfer from Conjugated Molecules to Structures, *Adv. Mater.* **25**(42), 6039–6049 (2013).

187. J. Wang, G. Wanf, X. Feng, T. Kitamura, Y. Kang, S. Yu, and Q. Qin, Hierarchical chirality transfer in the growth of Towel Gourd tendrils, *Sci. Rep.* **3**(1), 3102, (2013).

188. K. Schulgasser and A. Witztum, The hierarchy of chirality, *J. Theoretical Biol.* **230**(2), 281–288, (2004).

189. D. M. Ho and R. A. Pascal Jr., Decacyclene: A molecular propeller with helical crystals, *Chem. Mater.* **5**(9), 1358–1361 (1993).

190. D. Maillard and R. E. Prud'homme, Crystallization of ultrathin films of polylactides: From chain chirality to lamella curvature and twisting, *Macromolecules* **41**(5), 1705–1712 (2008).

191. K. L. Singfield, J. K. Hobbs, and A. Keller, Correlation between main chain chirality and crystal "twist" direction in polymer spherulites, *J. Cryst. Growth* **183**(4), 683–689 (1998).

192. J. Wang, C. Y. Li, S. Jin, X. Weng, R. M. Van Horn, M. J. Graham, W. Zhang, K. Jeong, F. W. Harris, B. Lotz, and S. Z. D. Cheng, Helical crystal assemblies in nonracemic chiral liquid crystalline polymers: Where chemistry and physics meet, *Ind. Eng. Chem. Res.* **49**(23), 11936–11947 (2010).

193. R. Plasson, D. K. Kondepudi, H. Bersini, A. Commeyras, and K. Asakura, Emergence of homochirality in far-from-equilibrium systems: Mechanisms and role in prebiotic chemistry, *Chirality* **19**(8), 589–600 (2007).

194. E. Nakouzi and O. Steinbock, Self-organization in precipitation reactions far from the equilibrium, *Sci. Adv.* **2**(8), e1601144 (2016).

195. S. Matsubara and H. Tamiaki, Photoactivated supramolecular assembly using "Caged Chlorophylls" for the generation of nanotubular self-aggregates having controllable lengths, *ACS Appl. Nano Mater.* **3**(2), 1841–1847 (2020).

196. A. Sorrenti, J. Leira-Iglesias, A. J. Markvoort, T. F. A. de Greef, and T. M. Hermans, Non-equilibrium supramolecular polymerization, *Chem. Soc. Rev.* **46**(18), 5476–5490 (2017).

197. K. Ruiz-Mirazo, C. Briones, and A. de la Escosura, Prebiotic systems chemistry: New perspectives for the origins of life, *Chem. Rev.* **114**(1), 285–366 (2014).

198. J. M. Ribó, D. Hochberg, J. Crusats, Z. El-Hachemi, and A. Moyano Spontaneous mirror symmetry breaking and origin of biological homochirality, *J. R. Soc. Interface* **14**(137), 20170699 (2017).

199. I. Danila, F. Riobé, F. Piron, J. Puigmartí-Luis, J. D. Wallis, M. Linares, H. Ågren, D. Beljonne, D. B. Amabilino, and N. Avarvari, Hierarchical chiral expression from the nano- to meso-scale in synthetic supramolecular helical fibers of a non-amphiphilic C_3-symmetrical π-functional molecule, *J. Am. Chem. Soc.* **133**(21), 8344–8353 (2011).

200. I. Danila, F. Pop, C. Escudero, L. N. Feldborg, J. Puigmartí-Luis, F. Riobé, N. Avarvari, and D. B. Amabilino Twists and turns in the hierarchical self-assembly pathways of a non-amphiphilic chiral supramolecular material, *Chem. Commun.* **48**(38), 4552–4554 (2012).

201. P. Xue, R. Lu, X. Yang, L. Zhao, D. Xu, Y. Liu, H. Zhang, H. Nomoto, M. Takafuji, and H. Ihara, Self-Assembly of a chiral lipid gelator controlled by solvent and speed of gelation, *Chem. Eur. J.* **15**(38), 9824–9835 (2009).

202. J. Kumar, T. Nakashima, H. Tsumatori, and T. Kawai, Circularly polarized luminescence in chiral aggregates: Dependence of morphology on luminescence dissymmetry, *J. Phys. Chem. Lett.* **5**(2), 316−321 (2014).

203. T. Kaseyama, S. Furumi, X. Zhang, K. Tanaka, and M. Takeuchi, Hierarchical assembly of a phthalhydrazide-functionalized helicene, *Angew. Chem. Int. Ed.* **50**(16), 3684–3687 (2011).

204. N. P. M. Huck, W. F. Jager, B. de Lange, and B. L. Feringa Dynamic control and amplification of molecular chirality by circular polarized light, *Science* **273**(5282), 1686–1688 (1996).

205. J.-Y. Kim, J. Yeom, G. Zhao, H. Calcaterra, J. Munn, P. Zhang, and N. Kotov, Assembly of gold nanoparticles into chiral superstructures driven by circularly polarized light, *J. Am. Chem. Soc.* **141**(30), 11739–11744 (2019).

206. R. Merindol and A. Walther, Materials learning from life: Concepts for active, adaptive and autonomous molecular systems, *Chem. Soc. Rev.* **46**(18), 5588–5619 (2017).

207. B. A. Grzybowski, K. Fitzner, J. Paczesny, and S. Granick, From dynamic self-assembly to networked chemical systems, *Chem. Soc. Rev.* **46**(18), 5647–5678 (2017).

208. B. Adhikari, J. Nanda, and A. Banerjee, Multi-component hydrogels from enantiomeric amino acid derivatives: Helical nanofibers, handedness and self-sorting, *Soft Matter* **7**(19), 8913–8922 (2011).

209. C. Colquhoun, E. R. Draper, E. G. B. Eden, B. N. Cattoz, K. L. Morris, L. Chem, T. O. McDonald, A. E. Terry, P. C. Griffiths, L. C. Serpell, and D. J. Adams, The effect of self-sorting and co assembly on the on the mechanical properties of low molecular weight hydrogels, *Nanoscale* **6**(22), 13719–13725 (2014).

210. G. C. Dizon, G. Atkinson, S. P. Argent, L. T. Santu, and D. B. Amabilino, Sustainable sorbitol-derived compounds for gelation of the full range of ethanol-water mixtures, *Soft Matter*, **16**(19), 4640–4654 (2020).

211. C. Thalacker and F. Würthner, Chiral perylene bisimide–melamine assemblies: Hydrogen bond-directed growth of helically stacked dyes with chiroptical properties, *Adv. Funct. Mater.* **12**(3), 209–218 (2002).

212. Y. Wang, D. Zhou, H. Lia R. Li, Y. Zhong, X. Sun, and X. Sun, Hydrogen-bonded supercoil self-assembly from achiral molecular components with light-driven supramolecular chirality, *J. Mater. Chem. C* **2**(31), 6402–6409 (2014).

213. R. L. Beingessner, Y. Fan, and H. Fenniri. Molecular and supramolecular chemistry of rosette nanotubes, *RSC Adv.* **6**(79), 75820–75838 (2016).

214. X. Zhu, Y. Jiang, D. Yang, L. Zhang, Y. Lia, and M. Liu, Homochiral nanotubes from heterochiral lipid mixtures: A shorter alkyl chain dominated chiral self-assembly, *Chem. Sci.* **10**(13), 3873–3880 (2019).

215. I. Destoop, H. Xu, C. Oliveras-González, E. Ghijsens, D. B. Amabilino, and S. De Feyter, 'Sergeants-and-Corporals' principle in chiral induction at an interface, *Chem. Commun.* **49**(68), 7477–7479 (2013).

216. N. Kameta and W. Ding, Supramolecular nanotube reactors for production of imine polymers with controlled conformation, size, and chirality, *Small* **15**, 1900682 (2019).

217. R. Oda, I. Huc, M. Schmutz, S. J. Candau, and F. C. MacKintosh, Tuning bilayer twist using chiral counterions, *Nature* **399**(6736), 566–569 (1999).

218. D. Berthier, T. Buffeteau, J.-M. Léger, R. Oda, and I. Huc, From chiral counterions to twisted membranes, *J. Am. Chem. Soc.* **124**(45), 13486–13494 (2002).

219. N. Ryu, Y. Okazaki, K. Hirai, M. Takafuji, S. Nagaoka, E. Pouget, H. Ihara, and R. Oda, Memorized chiral arrangement of gemini surfactant assemblies in nanometric hybrid organic–silica helices, *Chem. Commun.* **52**(34), 5800–5803 (2016).

220. Y. Okazaki, J. Cheng, D. Dedovets, G. Kemper, M.-H. Delville, M.-C. Durrieu, H. Ihara, M. Takafuji, E. Pouget, and R. Oda, Chiral colloids: Homogeneous suspension of individualized SiO_2 helical and twisted nanoribbons, *ACS Nano* **8**(7), 6863–6872 (2014).

221. Y. Okazaki, T. Buffeteau, E. Siurdyban, D. Talaga, N. Ryu, R. Yagi, E. Pouget, M. Takafuji, H. Ihara, and R. Oda, Direct observation of siloxane chirality on twisted and helical nanometric amorphous silica, *Nano Lett.* **16**(10), 6411–6415 (2016).

222. M. Attoui, E. Pouget, R. Oda, D. Talaga, G. Le Bourdon, T. Buffeteau, S. Nlate, Optically active polyoxometalate-based silica nanohelices: Induced chirality from inorganic nanohelices to achiral POM clusters, *Chem. Eur. J.* **24**(44), 11344–11353 (2018).

223. J. George, S. Kar, E. S. Anupriya, S. M. Somasundaran, A. D. Das, C. Sissa, A. Painelli, and K. G. Thomas, Chiral plasmons: Au nanoparticle assemblies on thermoresponsive organic templates, *ACS Nano* **13**(4), 4392–4401 (2019).

224. R. Xiong, J. Y. Luan, S. Kang, C. Ye, S. Singamaneni, and V. V. Tsukruk, Biopolymeric photonic structures: Design, fabrication, and emerging applications, *Chem. Soc. Rev.* **49**(3), 983–1031 (2020).
225. J. Liu, H. Zhou, W. Yang, and K. Ariga, Soft nanoarchitectonics for enantioselective biosensing, *Acc. Chem. Res.* **53**(3), 644–653 (2020).
226. K. Long, Y. Liu, Y. Li, and W. Wang, Self-assembly of trigonal building blocks into nanostructures: Molecular design and biomedical applications, *J. Mater. Chem. B* **8**(31), 6739–6752 (2020).
227. V. K. Praveen, B. Vedhanarayanan, A. Mal, R. K. Mishra, and A. Ajayaghosh, Self-Assembled extended π-systems for sensing and security applications, Acc. Chem. Res. **53**(2), 496–507 (2020).
228. R. Randazzo, M. Gaeta, C. M. A. Gangemi, M. E. Fragalà, R. Purrello, and A. D'Urso, Chiral recognition of L- and D- amino acid by porphyrin supramolecular aggregates, *Molecules* **24**(1), 84 (2019).
229. Y. Li, A. Hammoud, L. Bouteiller, and M. Raynal, Emergence of homochiral benzene-1,3,5-tricarboxamide helical assemblies and catalysts upon addition of an achiral monomer, *J. Am. Chem. Soc.* **142**(12), 5676–5688 (2020).
230. A. Sorrenti, R. Rodriguez-Trujillo, D. B. Amabilino, and J. Puigmarti-Luis, Milliseconds make the difference in the far-from-equilibrium self-assembly of supramolecular chiral nanostructures, *J. Am. Chem. Soc.* **138**(22), 6920–6923 (2016).
231. V. Cherpak, V. F. Korolovych, R. Geryak, T. Turiv, D. Nepal, J. Kelly, T. J. Bunning, O. D. Lavrentovich, W. T. Heller, and V. V. Tsukruk, Robust chiral organization of cellulose nanocrystals in capillary confinement, *Nano Lett.* **18**(11), 6770–6777 (2018).
232. G. Albano, F. Salerno, L. Portus, W. Porzio, L. A. Aronica, and L. Di Bari, Outstanding chiroptical features of this films of chiral oligothophenes, *ChemNanoMat* **4**(10), 1059–1070 (2018).
233. F. Zinna, C. Resta, M. Górecki, G. Pescitelli, L. Di Bari, T. Jávorfi, R. Hussain, and G. Siligardi, Circular dichroism imaging: Mapping the local supramolecular order in thin films of chiral functional polymers, *Macromolecules* **50**(5), 2054–2060, (2017).
234. G. Albano, M. Górecki, G. Pescitelli, L. Di Bari, T. Jávorfi, R. Hussain, and G. Siligardi, Electronic circular dichroism imaging (CDi) maps local aggregation modes in thin films of chiral oligothiophenes, *New J. Chem.* **43**(36), 14584–14592 (2019).
235. E. T. Pashuck and S. I. Stupp, Direct observation of morphological transformation from twisted ribbons into helical ribbons, *J. Am. Chem. Soc.* **132**(26), 8819–8821 (2010).
236. L. Zhang, J. Qin, S. Lin, Y. Li, B. Li, and Y. Yang, Aggregation-induced chirality: Twist and stacking handedness of the biphenylene groups of

n-$C_{12}H_{25}$O-BP-CO-Ala-Ala dipeptides, *Langmuir* **33**(41), 10951–10957 (2017).

237. W. Wu, W. Hu, G. Qian, H. Liao, X. Xu, and F. Berto, Mechanical design and multifunctional applications of chiral mechanical metamaterials: A review, *Mater. Des.* **180**, 107950 (2019).

238. Q. Lu, D. Qi, Y. Li, D. Xiao, and W. Wu, Impact energy absorption performances of ordinary and hierarchical chiral structures, *Thin-Walled Struct.* **140**(1), 495–505 (2019).

239. T. Frenzel, M. Kadie, and M. Wegener, Three-dimensional mechanical metamaterials with a twist, *Science* **358**(6366), 1072–1074 (2017).

240. M. Ikeda, Bioinspired supramolecular materials, *Bull. Chem. Soc. Jpn.* **86**(1), 10–24 (2011).

241. G. T. Zan and Q. S. Wu, Biomimetic and bioinspired synthesis of nanomaterials/nanostructures, *Adv. Mater.* **28**(11), 2099–2147 (2016).

242. N Bäumer, K. K. Karth, J. P. Palakkal, and G. Fernández, Morphology control in metallosupramolecular assemblies through solvent-induced steric demand, *Soft Matter* **16**(29), 6834–6840 (2020).

243. A. Ruiz-Carretero, N. R. Ávila Rovelo, S. Militzer, and P. J. Mésini, Hydrogen-bonded diketopyrrolopyrrole derivatives for energy-related applications, *J. Mater. Chem. A* **7**(41), 23451–23475 (2019).

244. S. Militzer, N. Nishimura, N. R. Ávila-Rovelo, W. Matsuda, D. Schwaller, P. J. Mésini, S. Seki, and A. Ruiz-Carretero, Impact of chirality on hydrogen-bonded supramolecular assemblies and photoconductivity of diketopyrrolopyrrole derivatives, *Chem. Eur. J.* **26**(44), 9998–10004 (2020).

245. Y. Yang, R. Correa da Costa, D.-M. Smilgies, A. J. Campbell, and M. J. Fuchter, Induction of circularly polarized electroluminescence from an achiral light-emitting polymer *via* a chiral small-molecule dopant, *Adv. Mater.* **25**(18), 2624–2628 (2013).

246. L. Đorđević, F. Arcudi, A. D'Urso, M. Cacioppo, N. Micali, T. Bürgi, R. Purrello, and M. Prato, Design principles of chiral carbon nanodots help convey chirality from molecular to nanoscale level, *Nat. Comm.* **9**, 3442 (2018).

247. J. R. Brandt, F. Salerno, and M. J. Fuchter, The added value of small-molecule chirality in technological applications, *Nature Rev. Chem.* **1**(6), UNSP0045 (2017).

248. M. Hua, H.-T. Feng, Y.-X. Y. Y.-S. Zheng, and B. Z. Tang Chiral AIEgens — Chiral recognition, CPL materials and other chiral applications, *Coord. Chem. Rev.* **416**(1), 213329 (2020).

Chapter 2

Insight on the Structure and Properties of Chiral Self-Assembled Natural Products*

Arie Zask[†,§], George Ellestad[‡,¶], and Nina Berova[‡,‖]

*†Department of Biological Sciences,
Columbia University, New York, NY, USA*

*‡Department of Chemistry, Columbia University,
New York, NY, USA*

§az2280@columbia.edu
¶gae2104@columbia.edu
‖ndb1@columbia.edu

This chapter reviews the remarkable structure and spectral properties of some selected examples of self-assembled natural products. These include carotenoids; flower pigments; hypericin, a potent, naturally occurring phenanthroperylene quinone photosensitizer; the seco-triterpenoid α-onocerin that self-assembles into remarkable flower-like structures in several solvents; and, finally, the light-harvesting antenna of green photosynthetic bacteria that make use of self-aggregates of bacteriochlorophylls for photosynthesis in environments with low light intensity.

---

*Dedicated to the memory of T. Silviu Balaban and in honor of his pioneering work on supramolecular chirality.

1. Introduction

In addition to the well-characterized assemblies of primary metabolites such as lipids, steroids, DNA, carbohydrates, and proteins, there are numerous examples of aggregates of so-called secondary metabolites or natural products and their spontaneous self-assembly. These aggregates play important biological roles in the producing organisms. In this chapter, we have provided a synopsis of selected examples of these remarkable self-assemblies that illustrate their unusual structure and the mechanisms used for their construction. In addition, we discuss the biological advantages that these complexes provide to the host organisms. These examples include carotenoids and flavonoid flower pigments, the quinone photosensitizer hypericin, the pentacyclic triterpene α-onocerin, and the light-harvesting antenna of green photosynthetic bacteria composed of bacteriochlorophylls. A number of biophysical techniques have been used to characterize these assemblies, but we emphasize here UV–Vis, ECD (electronic circular dichroism), and nuclear magnetic resonance (NMR) methods that provide important insights into the structures of these unusual aggregates.

1.1. *Driving forces for the self-aggregation of natural products*

These chiral natural product aggregates consist of assemblies of small molecular weight monomers that are self-assembled under conditions of thermodynamic equilibrium through kinetically stable non-covalent interactions. They are formed without any external protein control through several intermolecular attractive forces including electrostatic, van der Waals, hydrophobic, directional π–π stacking interactions, and metal coordination. An understanding of their structures and the mechanism by which they self-assemble is important to provide insight into their potential role in biological processes and their use for the treatment of a number of disease states.[1–5]

1.2. *Effect of self-assembly on biophysical properties*

Aggregated natural products provide, in general, a beneficial property to the producing organism, for example, a photoprotective role of carotenoid aggregates ensuring a wider spectral range of light absorption. In the case of chlorosome antenna complexes from green, photosynthetic bacteria, the light-harvesting abilities of aggregated chlorin pigments ensure sufficient light absorption to allow photosynthesis to occur in environments of very poor lighting. Homoassociation of hypericin, a naturally occurring hydroxyphenanthroperylene quinone, leads to a significant decrease in its absorption and fluorescent properties. The optical properties of hypericin appear to be beneficial to not only the producing plant but also as therapeutically useful antiviral, photodynamic, and antiproliferative agents. The pigments of flowers are composed of self-assembled homo- and heteroaggregates of anthocyanins and flavones in combination with metal ions and appended sugars, which account for the various shades and hues of flower color and provide color stability for the attraction of pollinating insects.

1.3. *Environmental factors of self-assembly*

The polarity of the cellular environment of the producing organism has a great influence on the self-assembly, especially with the more hydrophobic compounds like the carotenoids. Thus, more polar environments reinforce hydrophobic interactions and molecular crowding. The bacteriochlorophylls in chlorosomes are self-assembled through pigment–pigment interactions where again high concentrations of the monomers favor aggregation. Also, in the flower pigments, self-aggregation of the monomers occurs under conditions of molecular crowding and very specific π–π stacking of the aromatic chromophores. Polarity is also a big factor in the self-aggregation of the relatively water-insoluble hypericin under physiological conditions where self-association critically depends on the polarity of the solvent.

2. Selected Natural Product Aggregates

The natural products listed below have been chosen because of the unique structures of their aggregates. They are discussed in terms of the earlier-mentioned categories pertaining to the self-assembly process: the driving forces that encourage aggregation; the chiroptical and other spectral methods used to characterize the aggregates; how self-assembly relates to their biological properties; and the environmental conditions that promote aggregation.

2.1. *Carotenoids*

Carotenoids are important lipophilic plant pigments that aggregate in aqueous environments and play a critical role in photosynthetic systems and other biological functions.[6-8] They are responsible for the yellow hues exposed in plants after chlorophyll catabolism in the fall season each year. The structures of the monomeric carotenoids presented here (Figure 1) have been correlated with their self-aggregation to form assemblies with, surprisingly, very intense exciton-coupled Cotton effects compared to the monomers. UV–Vis and ECD spectroscopy have been used extensively to characterize the self-assembly of carotenoids and how the resulting induced supramolecular chirality plays a role in plant life. Indeed, it has been postulated that the reflected circularly polarized light of the chiral assemblies can be sensed by the compound eyes of insects, which in turn is important for the fertilization process. Spectroscopic studies by Zsila, Simonyi *et al.*[6,7] and a more recent analysis by Schweiggert *et al.*[8] have provided much insight into the type of aggregates — H cardpack or J head-to-tail or combinations of both — and their formation as a function of solvent, structure, and stereochemistry. For example, in (all-E)-lycopene (**1**) (Figure 1), a strong hypsochromic shift in the UV vibrational bands at 447, 474, and 507 nm and a loss of these bands and the appearance at 358 nm for the main absorption band in 20% acetone –80% water is indicative of coupled H-type aggregates (Figure 3).[6] These aggregates are stabilized by π–π interactions facilitated by the closely packed parallel and planar polyene chains (Figure 2). The γ-carotene (**2**) structure is characterized by one terminal ring that slightly distorts the planarity observed with lycopene. This also results in the loss of the vibrational bands and a shift of the main absorption band at 497 nm to 378 nm in the diluted sample, again characteristic of H-aggregates.

(All E)-β-carotene (**3**) contains two terminal cyclohexene rings (Figures 1 and 2) and in the dissolved state showed a smaller hypsochromic shift to 440, 461, and 498 nm in comparison with γ-carotene.[6] On dilution with water, no loss of the vibrational fine structure was observed, but a bathochromic shift to 440, 461, and 498 nm was observed that was previously associated with J-type aggregates (Figure 3). Clearly the presence of the two terminal rings forces apart the two polyene chains, thus diminishing their association. The UV–Vis for (9Z)-β-carotene (**4**) with the 9 double bond in the Z-configuration is essentially the same as for (all-E)-β-carotene. As might be expected based on the bulky terminal

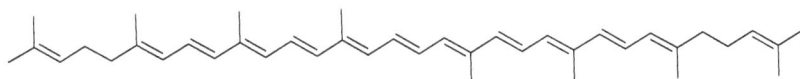

1. (All-E)-lycopene

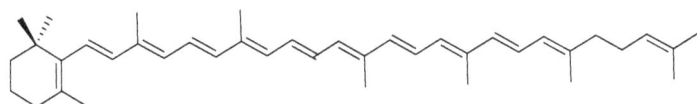

2. (All-E)-γ-carotene

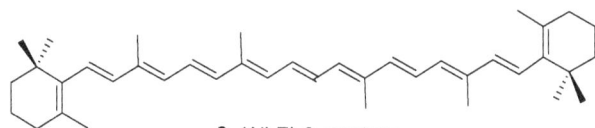

3. (All-E)-β-carotene

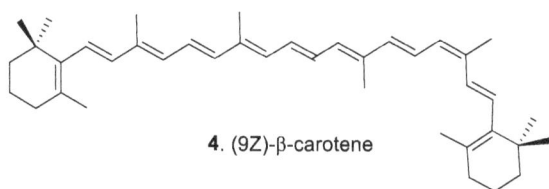

4. (9Z)-β-carotene

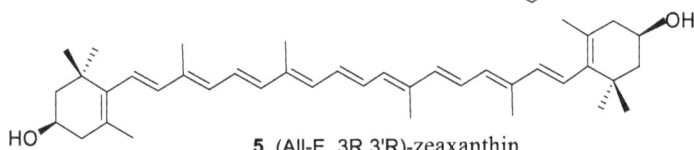

5. (All-E, 3R,3'R)-zeaxanthin

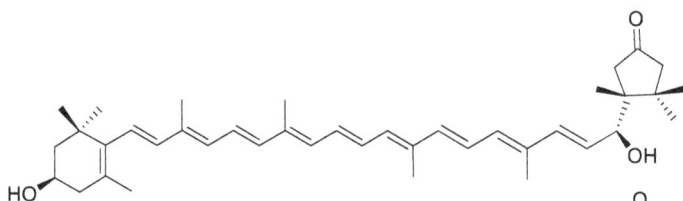

6. (6'S)-capsantholon

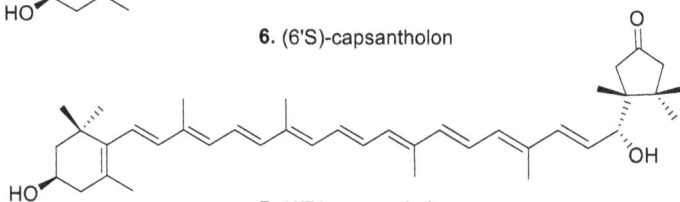

7. (6'R)-capsantholon

Figure 1.   Chemical structures of carotenoids discussed.

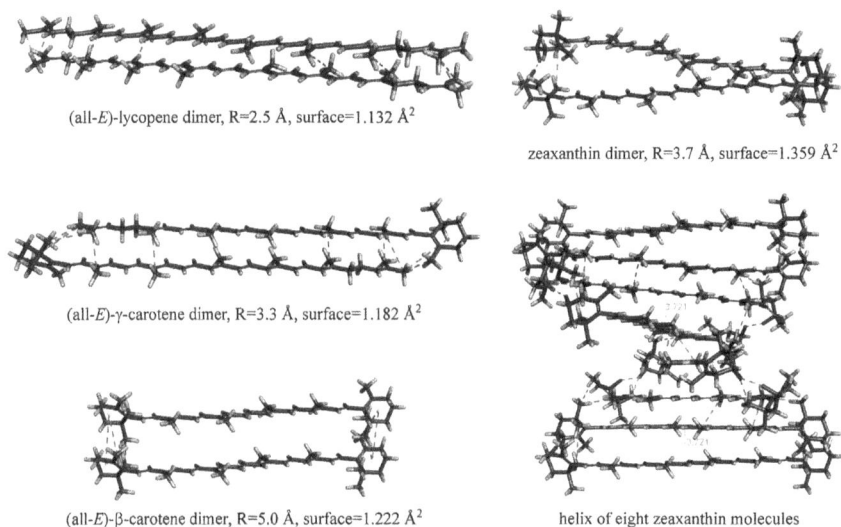

(all-*E*)-lycopene dimer, R=2.5 Å, surface=1.132 Å^2

zeaxanthin dimer, R=3.7 Å, surface=1.359 Å^2

(all-*E*)-γ-carotene dimer, R=3.3 Å, surface=1.182 Å^2

(all-*E*)-β-carotene dimer, R=5.0 Å, surface=1.222 Å^2

helix of eight zeaxanthin molecules

Figure 2.   Molecular models and solvent-accessible surface areas for selected carotenoid aggregates. Reprinted with permission from Ref. [8]. Copyright 2016, Elsevier.

rings, no pigment aggregation was observed based on its UV–Vis spectra as well as the absence of any ECD signals in the 300–700 nm region.

(All-E, 3R,3′R)-zeaxanthin (**5**) is a homodichiral carotenoid with one hydroxyl group on each of the cyclohexene rings at the end of the β-carotene molecule. Furthermore, the UV spectrum of the dissolved molecule showed vibrational bands at 429, 452, and 479 nm, similar to that of β-carotene. A large blue shift was observed on the addition of water as was an obvious loss of the vibrational fine structure. Upon addition of water to 80%, the UV maximum was shifted to 390 nm, and at this percentage of water a positive exciton CD couplet was observed with a positive Cotton effect at 397 nm followed by a negative one at 377 nm. An ECD signal inversion to a negative couplet was observed at 60% water.[8] Computational modeling of the larger aggregates seems to have resulted in a helical aggregate (Figure 2), which was supported by the observed bisignate exciton-coupled ECD spectrum mentioned earlier. This is apparently due to the formation of hydrogen bonds between the hydroxyl groups of the monomers.[6]

Simonyi *et al.*[7] studied molecular chirality transfer in the supramolecular formation of a number of carotenoid aggregates. In their study on

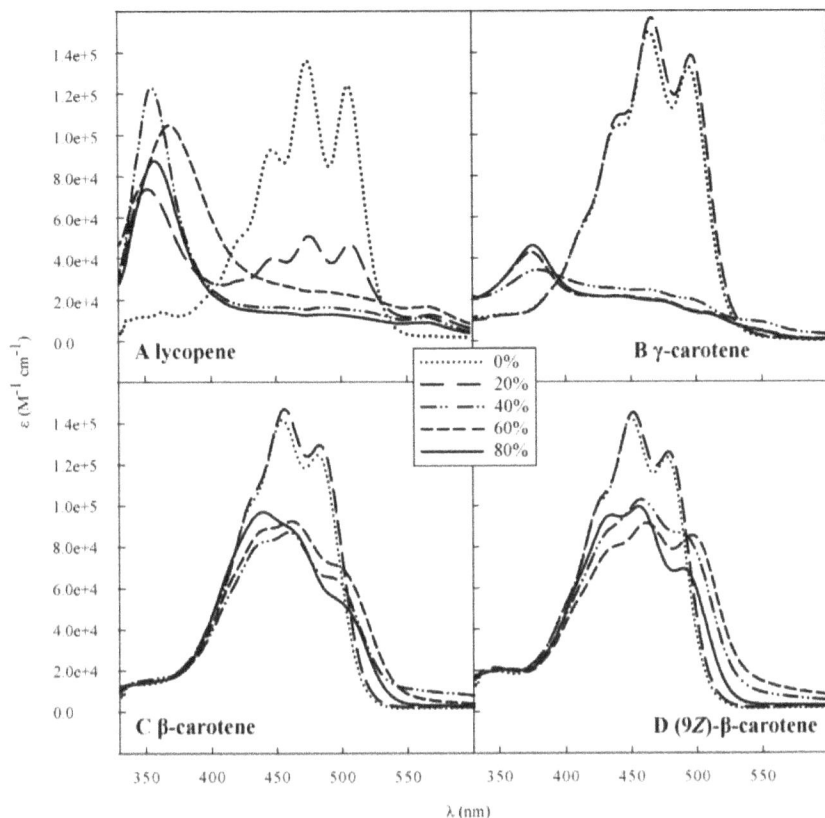

Figure 3.   UV–Vis spectral changes of lycopene, γ-carotene, β-carotene, and (9Z)-β-carotene upon dilution of acetone solutions with water. Reprinted with permission from Ref. [8]. Copyright 2016, Elsevier.

the self-assembly of (6'R)-capsantholon (**7**), it was shown that assembly proceeds as a function of the water/ethanol ratio (Figure 4). They observed that self-assembly began at an ethanol/water ratio of 1:1, as indicated by a new absorption at 377 nm compared to the monomer with gradual diminution of the vibrational bands of the monomer ascribed to a head-to-tail assembly. Furthermore, the intensity of the ECD Cotton effects increased some 100 times higher than that of the monomer in ethanol (base line above 300 nm) with the appearance of a newly formed negative exciton couplet at −372 and +388 nm (supramolecular chirality) along with residual contribution of head-to-tail J-type assemblies as suggested by the

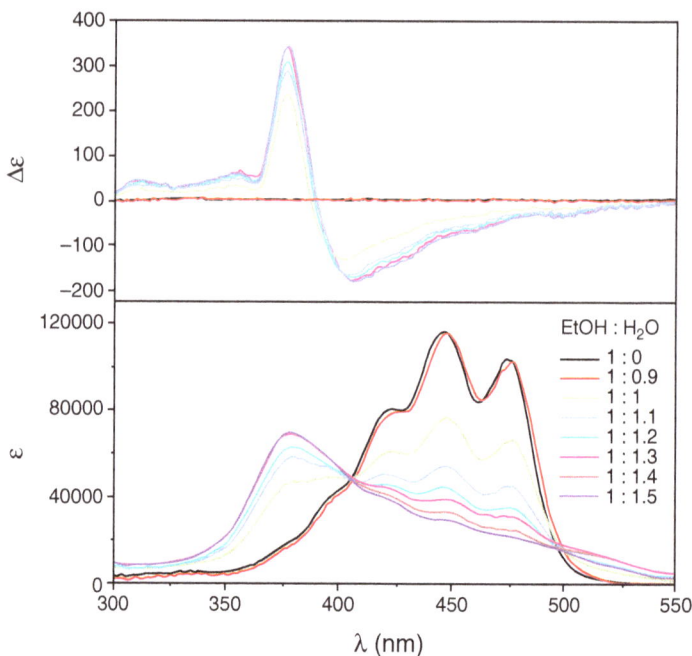

Figure 4.   ECD and UV–Vis spectral changes on dilution of (6'R)-capsantholon in ethanol with water. Reprinted with permission from Ref. [6]. Copyright 2003, Wiley.

red shifted tail seen in the UV spectrum. Obviously, the formation of a card-pack aggregate from a head-to-tail self-assembly is in contrast to that of the 6'S isomer (**6**), which, under the same conditions, shows no UV/ECD signature of card-pack aggregates.

The increase in intensity of Cotton effects for the 6'R isomer is clearly a result of the delocalization of the exciton energy over the closely stacked carotenoids in the aggregates.

2.2. *Hypericin*

Hypericin (**8**) is a photosensitizing pigment derived from plants of the genus *Hypericum* (Figure 5). In monomeric form, it exhibits a high quantum yield of fluorescence and photosensitizes the production of singlet oxygen. It functions as a defensive agent, deterring ingestion by animals and insects as a result of its toxicity on exposure to sunlight.

Hypericin is insoluble in neutral, aqueous environments and, under these conditions, forms nonfluorescent aggregates with significantly

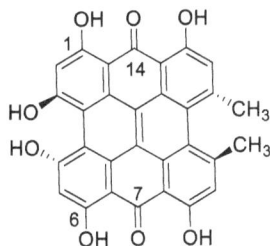

Figure 5. Structure and propeller conformation of the 7,14-dioxo hypericin (8) tautomer in the (P) enantiomeric form.

diminished photosensitizing ability.[9] In plants, hypericin is concentrated in granules (dark glands), likely relying on aggregation to diminish its photosensitizing properties. It was used as a folk medicine, and it has been extensively investigated for biological activities including antibacterial, antiviral, and antidepressant activities. Its photosensitizing properties has led to its potential application in cancer photodynamic therapy. However, its various biological activities and potential medical applications depend strongly on the properties of the aggregates, including their diffusion in aqueous environments.[10]

Hypericin is a phenanthroperylene quinone containing multiple hydroxyl and methyl groups and can exist as a number of tautomers, with the 7, 14-dioxo tautomer found to be most stable (Figure 5). It has C2 propeller geometry, is chiral, and separation of the enantiomers can be achieved by chiral High-performance liquid chromatography (HPLC).[11] Enantiomerization studies of separated hypericin enantiomers gave an experimental value for the racemization barrier of 97.1–99.6 kJ/mol, low enough that in biological systems hypericin exists as a racemate.[12]

Fluorescence measurements in dimethyl sulfoxide (DMSO)/water mixtures show that hypericin is monomeric at solutions up to 30% weight water.[9] At higher water concentrations, hypericin forms nonfluorescent aggregates containing at least four stacked, face-to-face hypericins or H-aggregates, maximizing the hydrophobic interaction of the aromatic cores (Figure 6).

These findings are based on nuclear Overhauser effects, appropriate NMR signals, and spin relaxation times. Figure 6 depicts a model showing homochiral or face-to-face P, P, P, P tetramers with neighboring molecules rotated 180° relative to each other and racemic P, P, M, M tetramers with neighboring P and M enantiomers rotated 90°. Measurements of the diffusion rates of hypericin in different water/DMSO mixtures showed that

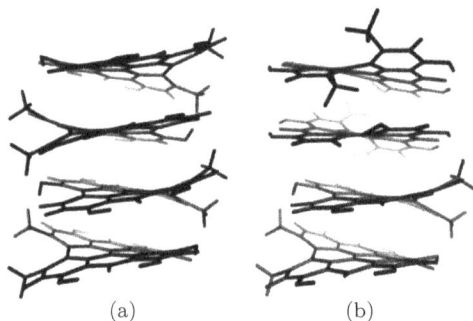

(a)   (b)

Figure 6.   Autodock 4 software calculated ball-and-stick models of two different stacked hypericin- tetramers: (a) homochiral P, P, P, P tetramer with adjacent molecules rotated about 180° from each other; (b) racemic P, P, M, M tetramer with adjacent P and M enantiomers rotated about 90° from each other. Reprinted with permission from Ref. [10]. Copyright 2011, American Chemical Society.

larger aggregates form with increasing amounts of water. In equal-weight mixtures of DMSO/water, the diffusion coefficient is 30% smaller than that calculated for the tetramer.[9]

Intracellular aggregation has been cited as the cause for decreased photocytotoxicity of hypericin in murine keratinocytes.[13] Fluorescent uptake time course studies show that with 5 uM hypericin, intracellular fluorescence intensity increased over time. However, cells treated with 50 uM hypericin showed a decrease of fluorescence intensity over time, indicative of aggregation. Photocytotoxicity in cells treated with 50 uM hypericin did not increase over time. In contrast, photocytotoxicity over time increased with 5 uM hypericin as expected, consistent with increasing intracellular concentrations of non-aggregated hypericin.

2.3. α-onocerin

The seco-triterpenpoid α-onocerin (**9**) has been observed to aggregate into remarkable flower-like and petal assemblies *via* the intermediacy of fibrillary networks with nano- to macromolecular diameters in various organic solvents such as o-dichlorobenzene and o-xylene as well as aqueous–organic solvents.[14] α-onocerin is extracted from the aerial part of *Lycopodium clavatum*. As seen in Figure 7, it has a unique C2-symmetric structure and is composed of two independent C15 trans-decalin units with two hydrophilic hydroxyl groupings at each end with a hydrophobic

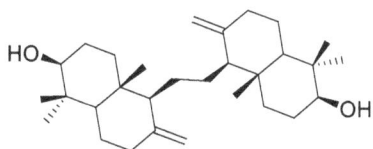

Figure 7. α-onocerin (**9**).

Figure 8. Examples of scanning electron microscopic images of flower-like assemblies of α-onocerin (**9**) in: (a–d) o-dichlorobenzene. Reprinted with permission from Ref. [14]. Copyright 2017, Wiley-VCH Verlag Gmbh&Vo.

interior connecting the two ends. The two hydrophobic trans-decalin units would aid in the aggregation process in aqueous environments.

Electron microscopy was used to study colloidal suspensions of α-onocerin in water-organic solvents such as o- and m-xylenes, bromobenzene, dodecanol, and DMSO-water. These mixtures provided unusual flower-like self-assemblies and fibrillary architectures, as seen in Figure 8.

Figure 9 shows a proposed molecular packing of α-onocerin, as viewed using wide-angle X-ray diffraction studies with dried gel samples from m-xylene and DMSO. The tilted lamellar-like oligomeric structure observed in Figure 8 with a length for each monomer of 1.51 nm is consistent with the FTIR OH stretching frequencies of the intermolecular

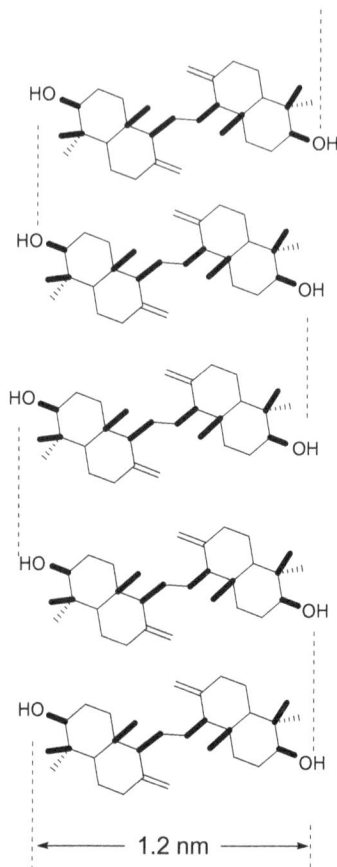

Figure 9.   Cartoon depiction of proposed molecular packing of α-onocerin (**9**) as suggested by wide-angle X-ray diffraction. Reprinted with permission from Ref. [14]. Copyright 2017, Wiley-VCH Verlag Gmbh & Vo.

hydrogen-bonded OH groupings at each end of the monomers. The diffraction pattern prepared from m-xylene shows a more ordered assembly than that from DMSO, although the diffraction data showed identical morphologies present in both samples.

These assemblies have been used to demonstrate the encapsulation and release of fluorophores such as Rho-8 in DMSO as a model compound for a drug delivery system. The adsorption of the fluorophore is facilitated by the porous nature and large surface area of the self-assemblies. These properties were also used to explain the encapsulation and release

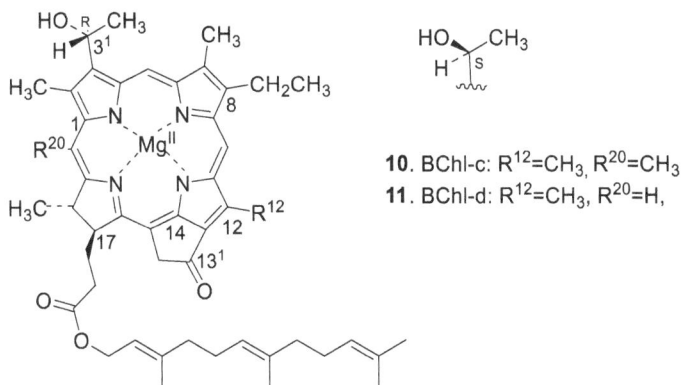

Figure 10. Molecular structures of chlorosomal bacteriochlorophylls c and d with the R and S configuration at C3[1].

of doxorubicin, an important anthracycline anticancer drug, thus demonstrating the potential utility for drug delivery applications.

2.4. *Self-assembly of bacteriochlorophylls into chlorosomes*

The light-harvesting antenna (chlorosomes) of green photosynthetic bacteria make use of self-aggregates of between 50,000 and 250,000 bacteriochlorophyll homologs with heterogeneity in size and substituents at positions 8 and 12 as well as in the long-chain alcohols, which esterify the 17-propionic acid.[15] Figure 10 shows the structures of bacteriochlorophylls c and d with the R and S configuration at C3[1] and esterified with farnesyl alcohol, and these are packed into rod-like aggregates forming chlorosomes (10–30 rods per chlorosome) depending on the species. Furthermore, bacteriochlorophylls exist as a mixture of diastereomers due to the R and S stereochemistry of the hydroxyethyl group at C3[1] and the S configuration at positions 17 and 18, which plays a key role in the self-assembly.[16] This diastereomeric control of self-assembly affects the optical properties of the oligomers that is not observed in their monomeric forms.

A recent review on this subject is given by Orf and Blankenship.[15] In contrast to the light-harvesting chlorophylls in plants, the bacteriochlorophylls from photosynthetic bacteria are self-assembled into supramolecular structures without the aid of proteins and are surrounded by glycol/phospholipid monolayers in an ellipsoid structure. Carotenoids are incorporated into the assemblies as energy acceptors for chlorosomes and triplet-state

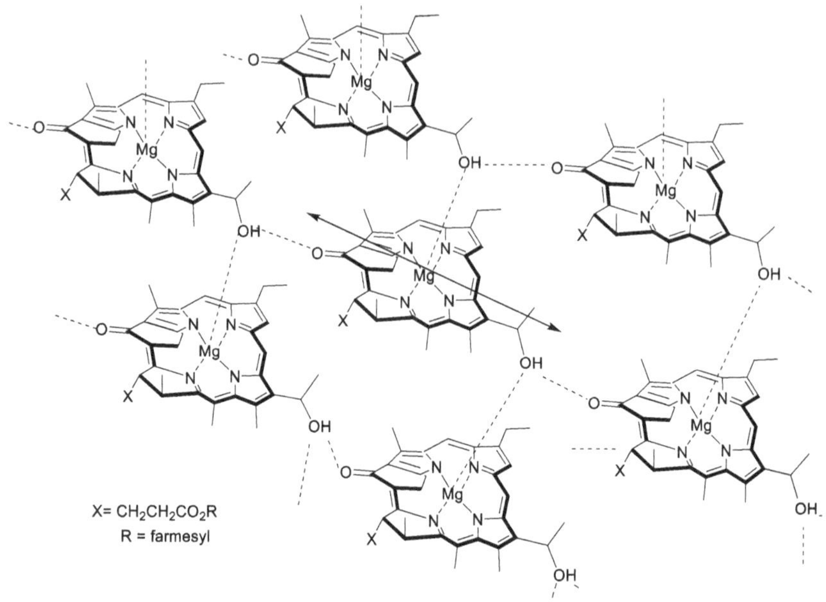

X= CH₂CH₂CO₂R
R = farmesyl

Figure 11.   Depiction of the hydrogen bond connectivities between bacteriochlorophyll monomer and the direction of the Qy axis. Reprinted with permission from Ref. [22]. Copyright 2005, Springer.

quenchers and aid in the self-assembly of the chlorosome. Quinones are also interspersed as excitation quenchers of oxygen to control photosynthesis. These supramolecular assemblies are driven by specific carbonyl–hydroxyl group bonding and coordination with Mg in addition to π–π interactions between the tetrapyrrole macrocycles structures and are wrapped into cylinders 5–10 nm in diameter (see Figure 11).

The growth mechanism of chlorosome formation has long intrigued researchers as to how the self-assembly of monomeric bacteriochlorophylls leads to supramolecular chlorosome structures with long rod-like elements. Balaban *et al.* developed a kinetic model for the self-aggregation of bacteriochlorophyll c by relating the appearance of the red-shifted absorption maximum at 739 nm to the aggregation state upon dilution of concentrated CH_2Cl_2 pigment solutions with n-hexane.[16] They found that aggregation did not occur instantaneously, but over time this maximum increased dramatically with pigment concentration and resulted in a sigmoidal growth curve response, similar to that of the growth of a crystal. The induction period appears to represent aggregation of monomers,

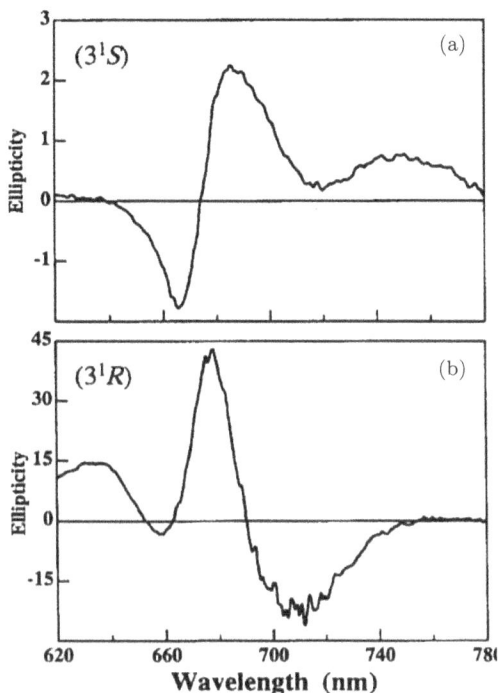

Figure 12. CD spectra of pure diastereomers (a) 3^1S and (b)3^1R bacteriochlorophyll c in dry methylene chloride. Reprinted with permission from Ref. [17]. Copyright 1995, Elsevier.

dimers, and lower aggregates in the 4–50 μM concentration range into larger chlorosomal aggregates with an absorption of 739 nm.

A number of studies have used ECD to analyze the structure of the stacks, and these have provided important insights into exciton coupling between adjacent pigment molecules and rod–rod interactions. Exciton coupling between the stacked monomers can be deduced from the ECD spectra in the near-infrared (IR) region, which is usually multisignate in shape. Balaban *et al.* studied the diastereoselective self-assembly of purified diastereomeric 3^1S and 3^1R bacteriochlorophyll c that indicted a strong dependency of solvent and the stereochemistry of the 3^1 hydroxyl groups.[17] Figure 12 shows the CD of pure diastereomeric 3^1S and 3^1R bacteriochlorophyll c. These spectra should be compared with the multisignate spectra of aggregates of various lengths and mixture of 3^1S and 3^1R hydroxyl groups shown in Figure 13. The structural heterogeneity of the aggregates, especially the lengths of the rod elements, results in wide variations in the intensity of the lobes as well as the order in which they appear. Prokhorenko *et al.* have shed much light on the heterogeneity of various ECD results and have found through model calculations that the spectrum changes shape from monomeric bacteriochlorophylls to ~45 chlorosome pigments per rod.[18] Thus, the difference in ECD results are largely due to the effective lengths of the exciton-coupled rods. Based on theoretical calculations (Frenkel exciton approach), they proposed, based on the earlier categorization studies of Griebenow *et al.*[19] with experimentally determined ECD curves from purified chlorosome preparations, that the shape of the spectra is composed of two contributions: (1) microscopic interchromophoric exciton coupling between adjacent monomer molecules showing a predominantly +/− sign going from short to long wavelengths and (2) macroscopic interrod exciton coupling between adjacent rods (−/+ from short to long wavelengths) using the absorbance at 750 nm (670 nm for a monomer) as a measure of aggregation. Thus, at very short lengths microscopic chirality is primarily a result of exciton coupling between adjacent pigment molecules with little contribution from rod–rod interactions, while macroscopic exciton coupling is due to interactions between the transition dipole moments of the monomers. As the rods lengthen, a Type 3 becomes apparent with the ECD shape becoming a combination of types 1 and 2 (−/+/−).

In Figure 13, a theoretical calculated ECD comparison between 10, 20, and 40 pigment molecules per stack is presented along with absorption, linear dichroism (LD), and convoluted stick spectra. The ECD spectra do not easily fit the above categorization but do illustrate the dramatic changes in the shape of the spectra on progressing from 10 to 40 molecules per stack.[18] Even at 10 molecules per stack, the shape of the ECD curve shows a mixture of types 1 and 2. For the 20 molecule per stack curve, the ECD curve is multi-signate with a −/+/−/+ signal; for the 40 molecule stack, the signal is −/+/− from short to long wavelengths. No further change in shape was observed with a higher number of molecules per stack. Thus, the calculated ECD curves explain the confusion with the early ECD curves from supposedly identical preparations of chlorosomes.

NMR has been especially useful in gaining structural insight into the molecular structure of chlorosomes and the role of R or S stereochemistry at C3[1] of the tetrapyrrole macrocycle. As mentioned above, the R and S stereochemistry plays a key role in chlorosome self-assembly, and the unique stacked structures shown in Figure 14 are important for extensive exciton delocalization in the excited state. For NMR studies, a structurally

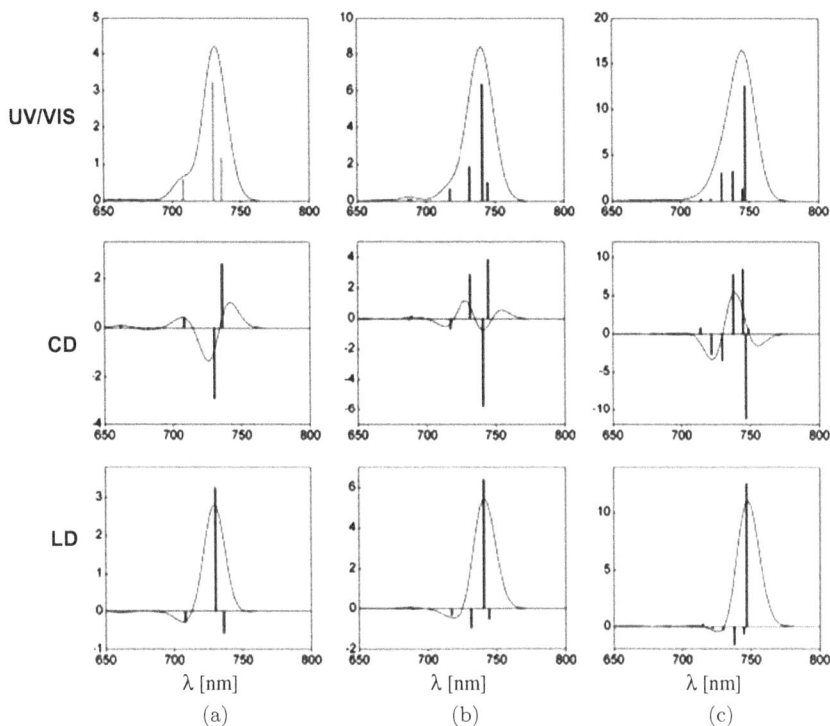

Figure 13.   Relation between rod length and shape of absorption, CD and LD spectra of aggregates of bacteriochlorophyll c (Figure 10); (a) 10, (b) 20, and (c) 40 molecules per stack calculated for a non-disordered system. The convoluted stick-spectra are normalized with a Gaussian profile of full-width at half maximum of 350 cm^{-1}. Reprinted from Ref. [18]. Copyright 2003, Biophysical Society.

homogenous (>95%) mutant was genetically prepared.[20] This was necessary due to the structural heterogeneity around the periphery of the tetra-pyrrole macrocycle in the wild-type chlorosomes[21] Thus, with a structurally homogenous and ^{13}C-enriched chlorosome, very informative two-dimensional ^{13}C–^{13}C, and ^{1}H–^{13}C solid state NMR revealed *syn-anti* dimerization with the C3 hydroxyethyl group in alternating diastereomeric conformations due to the mixture of R or S at C3^{1}. This accommodation of heterogeneity in the C3 side chain was proposed by the authors for evolutionary optimization of the enhancement of the organism's light-harvesting properties without a protein template. As seen in Figure 14, the *syn* form has the OHs ligated *cis* to the farnesyl ester side chain and *anti* in the opposite or *trans* orientation. Doubling of specific ^{1}H and ^{13}C signals in this mutant

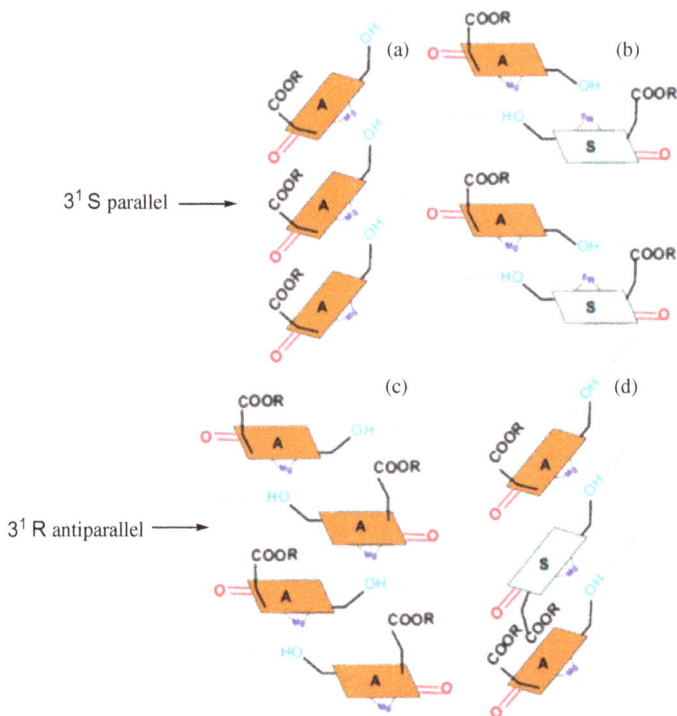

Figure 14.   Cartoon models of four possible structures of bacteriochlorophyll building blocks: (a) Parallel-stack model. (b) Piggy-back dimer model. (c) Antiparallel monomer stack. (d) Syn-anti monomer stack. As seen in the figure, stacking occurs *via* C3' OH-Mg coordination in all of the models. The parallel and antiparallel designations on the left of the image refer to *in vitro* oligomerization results of epimerically pure R and S diastereomers. Reprinted with permission from Ref. [20]. Copyright 2009, National Academy of Science of the United States of America.

as well as for the wild type led to two correlation networks of approximately 1:1 ratio. These were assigned to a parallel stack ((a) in Figure 14) for the wild-type or *syn-anti* stack for the mutant ((d) in Figure 14). This led to the rejection of the piggy-back dimer (b) and the antiparallel structure (c) due to the mixing of *syn-* or *anti-*coordination.

As seen in Figure 15, the hydrogen-bonding between the C13^1 and C3^1 hydroxyl groups directs the pigments transition dipole orientation such that the exciton delocalization can move up or down the length of the oligomer, thus ensuring rapid energy transfer in the system, which is critical for poor light environments. NMR studies on the 8-ethyl, 12-methyl

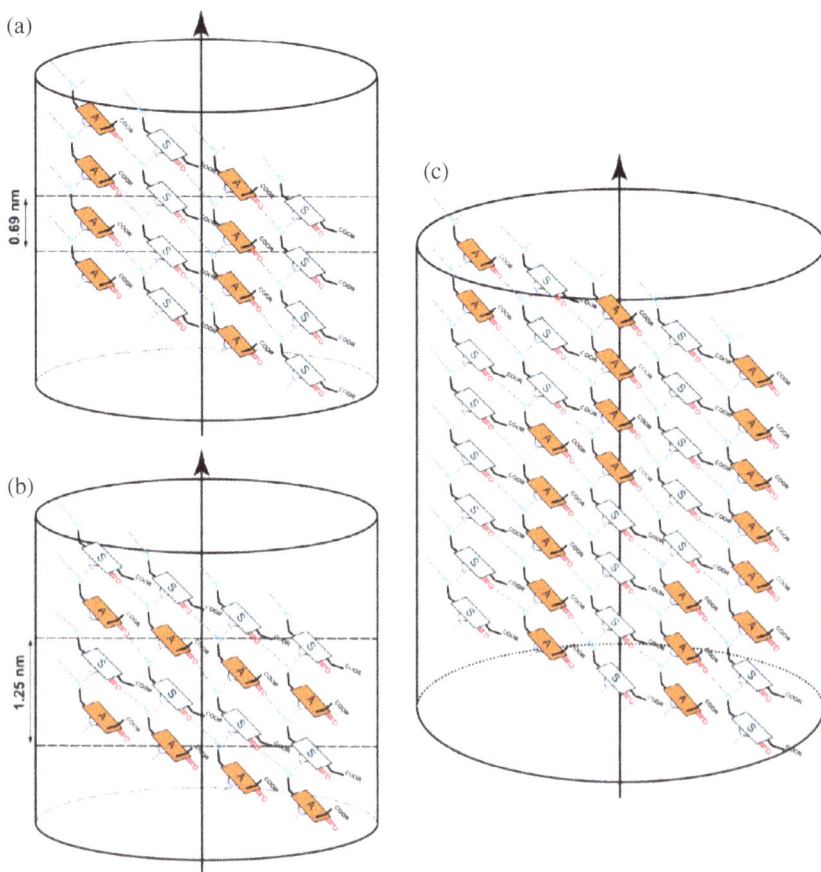

Figure 15.   Cartoon depictions of a section of (a), the chlorosome mutant used in the NMR studies; and (b), for the wild type. Domains can be joined together by *syn-anti* interfaces to introduce heterogeneity as in (c). The arrows indicate the long-axis of the chlorosome. **A** indicates an *anti* orientation and **S** a *syn* orientation of the monomers. Reprinted with permission from Ref. [21]. Copyright (2012), American Chemical Society.

bacteriochlorophyll c mutant revealed that *syn-anti* parallel stacks can introduce heterogeneity in the aggregates by connecting *syn-* and *anti-*coordinated domains with helical regions as indicated in Figure 15. The mixture of all *syn* and all *anti* parallel stacks in the mutant is apparently not optimal for light harvesting, and subsequent evolutionary development gave rise to tubular supramolecular aggregates with a dimeric repeating unit of a *syn-anti*-dimer in the direction of the long axis, as shown in Figure 15(c). Thus, this unique orientation provides for efficient long-range energy-harvesting capabilities.

It should be mentioned that Miyatake and Tamiaki studied the *in vitro* aggregation of epimerically pure R and S C3[1] diastereomers with a long-chain alcohol esterified at the propionic side chain.[22] They observed that the R epimer gave the antiparallel oligomer, whereas the S epimer gave preferentially the parallel oligomer (see Figure 14). The preference of the R epimer for the antiparallel stack is in contrast to the above-mentioned NMR studies on wild-type chlorosomes, which ruled out the antiparallel B stacks. They also found that the addition of small amounts of the 3[1] S epimer to R dimeric isomers converted the latter from antiparallel oligomers to parallel oligomers. But it turns out that both R and S C3[1] diastereomers are necessary for making the rod-shaped supramolecular aggregates in the wild-type chlorosome.

2.5. *Flower pigments*

Flower pigments are some of the best examples of chirality in supramolecular assemblies in which the glycosides of anthocyanins are stacked with co-pigment glycosides of flavones complexed with metal cations through phenolic groups. The co-pigmentation is necessary to protect and stabilize the color of the anhydro base anion of the anthocyanins from the weakly acid pH of the plant vacuoles (Figure 16). Recent reviews by Yoshida *et al.* provide a detailed summary of the chemical mechanisms of color development in these complexes.[23,24] The metal phenolic complexes make up the center of the complex and help form unique assemblies that correlate with a certain color.[23] The attached peripheral sugars direct the absolute stereochemistry of the self-assembled monomers. ECD has been an important spectroscopic tool in elucidating the stackings of the various complexes.

In Figure 17, Yoshida *et al.*, summarized three different ways polyacylated anthocyanins (e.g., malonylawobanin in Figure 18) are self-assembled into ordered intra- and intermolecular stacks, which can be characterized by ECD.[24] In these aggregates, the aromatic acyl groups, e.g., *p*-coumaroyl, caffeoyl, feruloyl, and *p*-hydroxybenzoyl moieties, primarily in bluish flowers, stack on either side of an anthocyanin molecule and enhance blue color development.

The cartoon structure of the deep blue pigment commelinin from *Commelina communis* based on a stereoscopic wire model from the

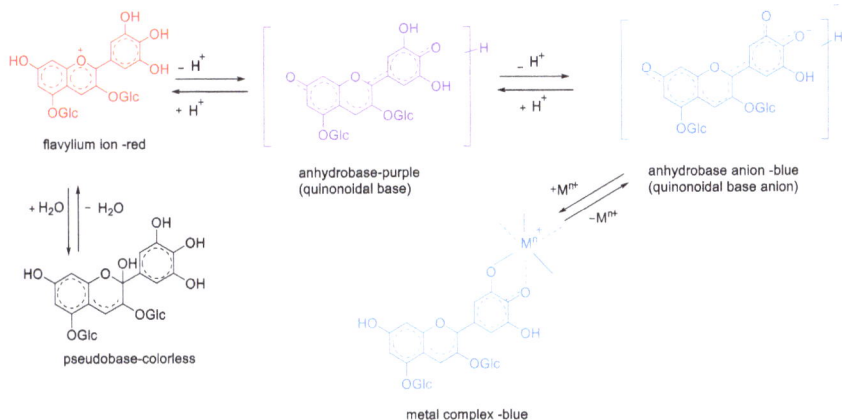

Figure 16. Structural changes in color in the anthocyanin chromophore as a function of pH. Reprinted with permission from Ref. [23]. John Wiley & Sons, Ltd.

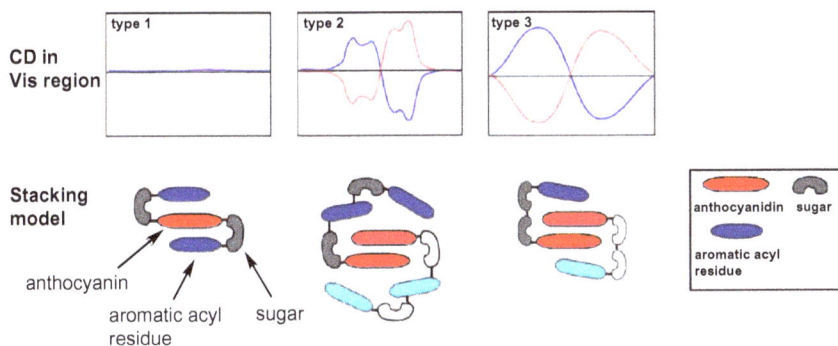

Figure 17. Schematic structures of the various stackings observed in bluish aromatic acylated anthocyanin-containing pigments as proposed by CD in the visible region. Type 1: intramolecular sandwich-type stacking, Type 2: pigments with intra- and inter-molecular stacking, and Type 3: Nested inter-molecular stacking. Red-orange: anthocyanidins, blue: aromatic acyl, grey or white: sugars. Reprinted with permission from Ref. [24]. Copyright 2009, The Royal Society of Chemistry.

seminal X-ray analysis from Goto's group[25,26] is shown in Figure 18 and depicts the intercalation of the quinonoidal form of anthocyanins (red) with flavones (blue) to stabilize the color and is an excellent example of this type of self-assembly.[27] This success really broke open flower pigment

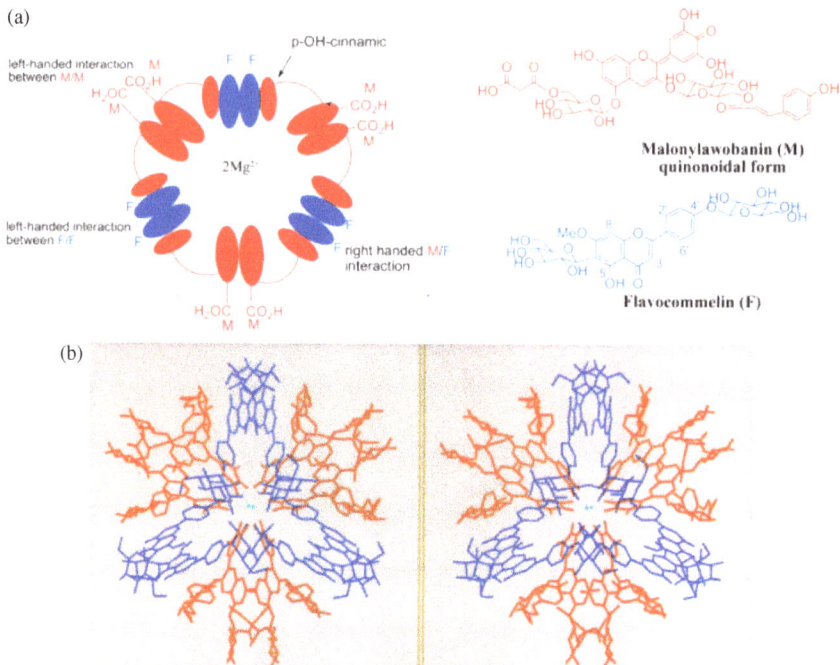

Figure 18. (a) Cartoon structure of commelinin. (b) Stereoscopic depiction of wire models of Cd-commelinin used for X-ray analysis. The anthocyanins are colored red, the flavones blue, and the two Cds are colored green dots in the center of the complex. Reprinted with permission from Ref. [27]. Copyright 2006, John Wiley & Sons, Ltd.

research after many decades of failures to explain the structures of these complexes.

Similar structures are observed with other pigments such as protocyanin from the cornflower (see Figure 19).[27,28] The self-assembly takes place under conditions of molecular crowding and specific π−π stacking interactions between the aromatic rings in addition to intermolecular hydrogen bonding between the attached sugars. Yoshida *et al.* carried out a number of synthetic reconstruction studies between anthocyanin and flavone components and, for example, in the case of malonylawobanin and flavocommelin (Figure 18), showed that complexation occurs only with D sugars of flavocommelin and is concentration dependent.[23]

The CD spectrum (Figure 20) of commelinin[24] is obviously a combination of more than one exciton couplet. The CD spectra of the complex shows strong, negative, superimposed exciton-type Cotton effects

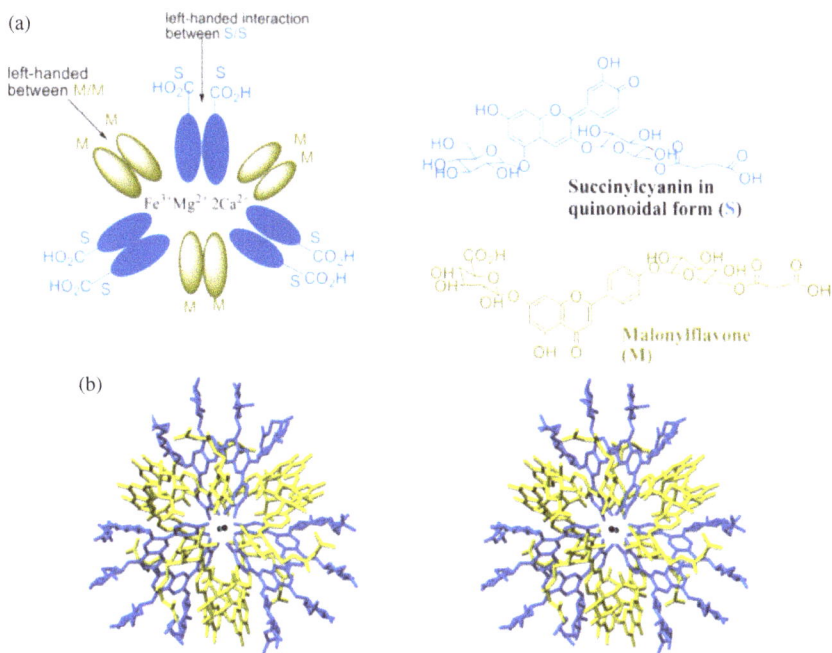

Figure 19. (a) Cartoon structure of protocyanin from blue petals of the cornflower, *Centurea cyanus*. (b) Stereoscopic view along the 3-fold axis of a wire model of protocyanin. The metals are depicted as small spheres in the center of the complex. Similar to the commelinin complex, the anthocyanin and flavone homodimers in protocyanin are assembled in a left-handed stack as evidence by a strong negative exciton couplet at −636 nm and +597 nm the in the CD spectrum. Reprinted with permission from Ref. [27]. Copyright 2006, John Wiley & Sons, Ltd.

around 600 nm in the visible region indicative of a left-handed screw orientation between the anthocyanins, consistent with the X-ray and NMR studies.[25,26]

The X-ray structure also indicated a left-handed interaction between the two flavones and a right-handed, or clockwise, relationship between the anthocyanin and flavone. The flavone co-pigment itself is essentially insoluble in water but becomes soluble when mixed with the anthocyanin and Mg.

3. Conclusions

In summary, the driving force for the self-assembled aggregates discussed in the above examples is a result of a number of interrelated components: the

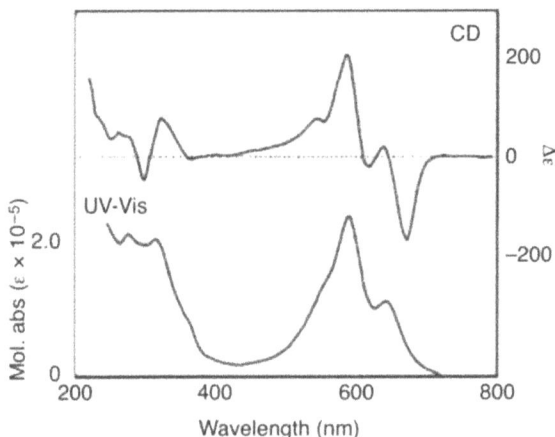

Figure 20.   UV–Vis and CD spectra of the deep blue flower of commelinin at 50 μM in aqueous solution at pH 5.0. Reprinted with permission from Ref. [27]. Copyright 2006, John Wiley & Sons, Ltd.

combination of the aqueous polarity of the environment, the mainly hydrophobic surface areas of the monomers, and the molecular crowding brought about by the polarity-induced hydrophobic binding between the monomers. The structures of the aggregates are controlled by a combination of factors including hydrogen bonding, π–π interactions, and metal complexation that together give rise to the remarkable structures that appear to have evolved to support the biological needs of the producing organism. These structures include precisely stacked and exciton-coupled bacteriochlorophylls for photosynthesis to convert sunlight into chemical energy in light-poor environments, carotenoids for the protection of the organism against damaging reactive oxygen species, the unique photochemistry of hypericin for protection against predators and antiviral and cancer properties, the porous structures of the aggregated triterpenoid α-onocerin for potential drug delivery, and the exciton-coupled flower pigment complexes that provide the beautiful colors that may play a role in insect pollination. The use of NMR, UV–Vis, CD and other biophysical spectroscopies have been critical in unraveling the structures of these complexes and their role in the producing organism. CD has been especially useful in revealing exciton coupling in the finely tuned self-assembled aggregates.

References

1. J.-M. Lehn, Supramolecular chemistry: Receptors, catalysts, and carriers, *Science* **227**, 849–856 (1985).
2. J.-M. Lehn, Perspectives in supramolecular chemistry−from molecular recognition towards molecular information processing and self-organization, *Angew. Chem. Int. Ed. Engl.* **29**, 1304–1319 (1990).
3. J.-M. Lehn, Supramolecular chemistry, *Science* **260**, 1762–1763 (1993).
4. D. N. Reinhoudt and M. Crego-calama, synthesis beyond the molecule, *Science* **295**, 2403–2407 (2002).
5. J.-M. Lehn, Perspectives in chemistry-aspects of adaptive chemistry and materials, *Angew. Chem. Int. Ed.* **54**, 3276–3289 (2015).
6. M. Simonyi, Z. Bikádi, F. Zsila, and J. Deli, Supramolecular exciton chirality of carotenoid aggregates, *Chirality* **15**, 680–698 (2003).
7. F. Zsila, J. Deli, Z. Bikádi, and M. Simonyi, supramolecular assemblies of carotenoids, *Chirality* **13**, 739–744 (2001).
8. J. Hempel, C. N. Schädle, S. Lepthin, R. Carle, and R. M. Schweiggert, structure related aggregation behavior of carotenoids and carotenoid esters, *J. Photochem. And Photobiol A: Chemistry* **317**, 161–174 (2016).
9. H. Falk, From the photosensitizer hypericin to the photoreceptor stentorin the chemistry of phenanthroperylene quinones, *Angew. Chem. Int. Ed.* **38**, 3116–3136 (1999).
10. G. Bánó, J Staničová, D. Jancura, J. Marek, M. Bánó, J. Uličný, A. Strejčková, and P. Miškovský, On the diffusion of hypericin in dimethylsulfoxide/water mixtures — the effect of aggregation, *J. Phys. Chem. B.* **115**, 2417–2423 (2011).
11. L. Sanders, M. Halder, T. L. Xiao, J. Ding, D. W. Armstrong, and J. W. Petrich, The separation of hypericin's enantiomers and their photophysics in chiral environments, *Photochem. Photobiol.* **81**, 183–186 (2005).
12. A. Ciogli, W. Bicker, and W. Lindner, Determination of enantiomerization barriers of hypericin and pseudohypericin by dynamic high-performance liquid chromatography on immobilized polysaccharide-type chiral stationary phases and off-column racemization experiments, *Chirality* **22**, 463–471 (2010).
13. T. Theodossiou, M. D. Spiro, J. Jacobson, J. S. Hothersall, and A. J. MacRobert, Evidence for intracellular aggregation of hypericin and the impact on its photocytotoxicity in PAM 212 murine keratinocytes, *Photochem. Photobiol.* **80**, 438–443 (2004).
14. B. G. Bag, S. N. Hasan, P. Pongpamorn, and N. Thasana, First hierarchical self-assembly of a seco-triterpenoid α-onocerin yielding supramolecular architectures, *ChemistrySelect* **2**, 6650–6657 (2017).
15. G. S. Orf and R. E. Blankenship, Chlorosome antenna complexes from green photosynthetic bacteria, *Photosynth. Res.* **116**, 315–331(2013).

16. T. S. Balaban, J, Leitich, A. R. Holzwarth, and K. Schaffner, Autocatalyzed self-aggregation of (3^1R)-[Et,Et]bactereocholorphyll c_F molecules in nonpolar solvents. Analysis of the kinetics, *J. Phys. Chem. B.* **104**, 1362–1372 (2000).

17. T. S. Balaban, A. R. Holzwarth, and K. Schaffner, Circular dichroism study on the diastereoselective self-assembly of bacteriochlorophyll c_s, *J. Mol. Struct.* **349**, 183–186 (1995).

18. V. I. Prokhorenko, D. S. Steensgaard, and A. R. Holzwarth, Exciton theory for supramolecular chlorosomal aggregates: 1. Aggregated size dependence of the linear spectra, *Biophys. J.* **85**, 3173–3186 (2003).

19. K. Griebenow, A. R. Holzwarth, F. van Mourik, and R. van Grondelle, Pigment organization and energy transfer in green bacteria. 2. Circular and linear dichroism spectra of protein-containing and protein-free chlorosomes isolated from *Chloroflexus aurantiacus* strain Ok-70-fl, *Biochim. Biophys. Acta.* **1058**, 194–202 (1991).

20. S. Ganapathy, G. T. Oostergetel, P. K. Wawrzyniak, M. Reus, A. G. M. Chew, F. Buda, E. J. Boekema, D. A. Bryant, A. R. Holzwarth, and H. J. M. de Groot, Alternating *syn-anti* bacteriochlorophylls form concentric helical nanotubes in chlorosomes, *PNAS.* **106**, 8525–8530 (2009).

21. S. Ganapathy, G. T. Oostergetel, M. Reus, Y. Tsukatani, A. G. M. Chew, F. Buda, D. A. Bryant, A. R. Holzwarth, and H. J. M. de Groot, Structural variability in wild-type and *bchQ bchR* mutant chlorosomes of the green sulfur bacterium, *Chlorobaculum tepidum*, *Biochem.* **51**, 4488–4498 (2012).

22. T. Miyatake and H. Tamiaki, Self-aggregates of bacteriochlophylls-*c*, *d* and *e* in a light-harvesting antenna system of green photosynthetic bacteria: Effect of stereochemistry at the chiral 3-(1-hydroxyethyl) group on the supramolecular arrangement of chlorophyllous pigments, *J. Photochem. and Photobiol. C: Photochem. Rev.* **6**, 89–107 (2005).

23. K. Yoshida, K-i. Oyama, and T. Kondo, Chemistry of Flavonoids in Color Development, in *Recent Advances in Polyphenol Research*, vol. 3, First Edition, Vèronique Cheyner, Pascale Sarni-Manchado, and Stéphane Quideau (Eds.), John Wiley & Sons, Ltd., Chichester, United Kingdom, (2012).

24. K. Yoshida, M. Mori, and T. Kondo, Blue color development by anthocyanins: From chemical structure to cell physiology, *Nat. Prod. Rep.* **26**, 884–915 (2009).

25. T. Goto and T. Kondo, Structure and molecular stacking of anthocyanins-flower color variation, *Angew. Chem. Int. Ed. Engl.* **30**, 17–33 (1991).

26. T. Kondo, K. Yoshida, A. Nakagawa, T. Kawai, H. Tamura, and T. Goto, Structural basis of blue-color development in flower petals from *Commelina communis*, *Nature* **358**, 515–518 (1992).

27. G. A. Ellestad, Structure and chiroptical properties of supramolecular flower pigments, *Chirality* **18**, 134–144 (2006).

28. M. Shiono, N. Matsugaki, and K. Takeda, Structure of the blue cornflower pigment, *Nature* **436**, 791 (2005).

Chapter 3

Chirogenesis in Supramolecular Systems

Kuppusamy Kanagaraj[*,‡], Cheng Yang[*,§],
and Victor Borovkov[†,¶]

*Key Laboratory of Green Chemistry and Technology,
College of Chemistry, State Key Laboratory of Biotherapy,
West China Medical Center, and Healthy Food Evaluation Research
Center, Sichuan University, Chengdu, China*

†*Department of Chemistry and Biotechnology,
School of Science, Tallinn University of Technology,
Akadeemia tee 15, 12618 Tallinn, Estonia*

‡*kanagaraj195@gmail.com*
§*yangchengyc@scu.edu.cn*
¶*victor.borovkov@taltech.ee*

Chirogenesis is a unique process occurring in natural systems and an interdisciplinary emerging research field of science in the recent era. The creation of chirogenesis effects in the supramolecular systems *via* non-covalent interactions is the most fascinating, diverse, and adaptable, which offers new and exciting possibilities/applications for the chirogenic hosts and assembly of materials with unprecedented features. Here, the present results mainly highlight the chirality generation, and its chirogenic effects in the selected methods and strategies are discussed. More specifically, the fundamental methods/strategies and challenges for the design and construction of supramolecular chirogenic systems are discussed, and we surveyed these important systems in the major chirogenic applications like chiral induction, sensing/recognition, discrimination, and photochirogenesis processes.

1. Introduction

Chirality is one of the most fundamental principles of nature, being the heart of chemical sciences and describing the ability of any object to exist as a pair of non-superimposable mirror images. The chirality of a single molecule is generated from its stereogenic center, which is originated/positioned at an atom(s) or any geometrical point in the chemical structure. These chiral centers are called stereogenic elements, which determine whether the given structure is chiral or not.[1] Chirality is not only applicable to the covalently linked molecules with well-defined configuration and conformation but can be also applied to the non-covalently bonded systems driven by supramolecular interactions with conformational flexibility and transient molecular conformations. Besides a single stereogenic center, there are many other types of alternative chiral elements defining molecular chirality (Figure 1).[2]

Among these types, a single molecular stereogenic element such as a center, axis, or plane is not always sufficient to describe the chirality of supramolecules like rotaxanes, catenanes, fullerenes, cavitands, or capsular assemblies and self-assemblies (helical).[2,3] Discovery of these new (supra)molecular arrangements (in the form of helix and curvature) resulted in the development of new independent stereo-sensors being able to differentiate its absolute configuration in the case of both inherent and induced chiral systems. Unique properties of these new supramolecular systems are significantly emerging and being applied in different branches of chemistry and materials science in recent years.[4]

As the chirality is essentially unique and inherent in biological systems which are involved in the functioning of various biological activities in living systems, it is apparent that this phenomenon exists in diverse forms and various scales from molecular to supramolecular levels. Therefore, all actions with chirality (inherent or induced), such as generation, amplification, monitoring, and controlling, are a special type of art and science for chemists, as in small and in highly complexed molecular structures. In this respect, the supramolecular approach is an intellectually challenging exercise resulting in a new interdisciplinary area of chemistry,

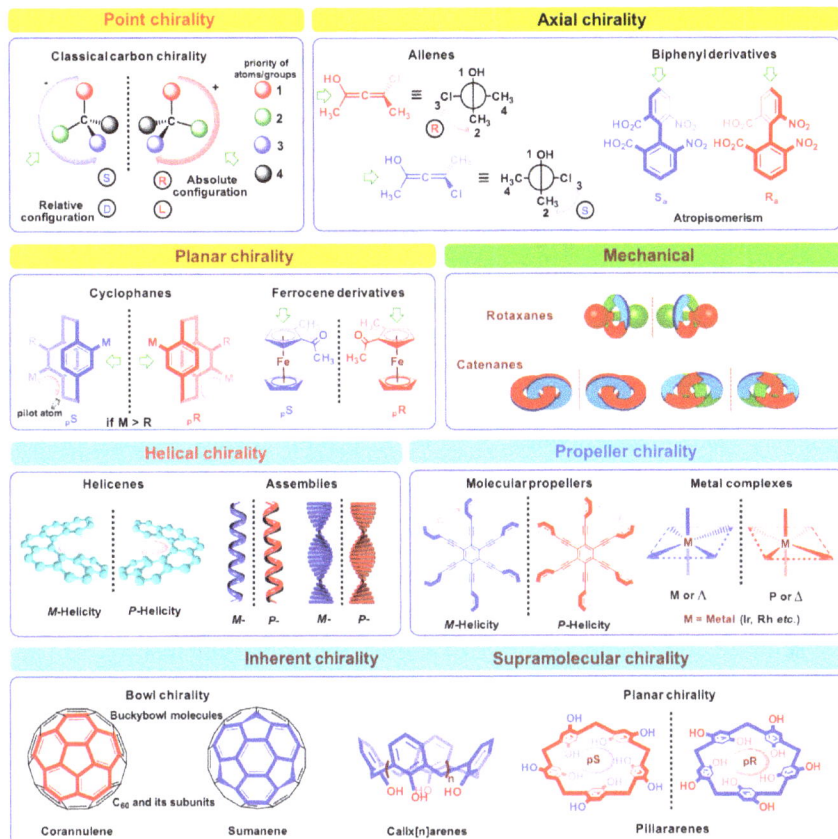

Figure 1. Different types of chirality and representative examples of different stereogenic elements in chemical structures.

which is called *supramolecular chirogenesis*, where the chiral/asymmetric information transfers within multi-component systems *via* non-covalent interactions such as hydrogen bonding, van der Waals contacts, π–π binding, hydrophobic, steric, and attractive or repulsive electrostatic interactions. This is a fascinating process in host–guest or self-assembled systems, with the primary/main mode of chiral communication occurring either independently or cooperatively to produce the enantiopure/homochiral molecules (in the case of chiral/asymmetric catalysis) and/or chiral structures (in the case of chiral sensing/recognition). The chirogenesis occurs as from host to guest or *vice versa*, while various external factors and stimuli (i.e., light, pH, solvent, additives, etc.) may control this whole process, with the host or guest or its combined structure rendering them to be chiral. In the case of chirogenic self-assembled structures, the chirality is induced and/or amplified/modulated through the self-assembly process, which is often applied in catalysis, sensing, nanotechnology, biomedical and material applications.[4a,5]

In this respect, chirogenesis is an important and peculiar process involved in supramolecular systems, which was an elegantly realized system as it mimics many processes in biological systems. There are, however, numerous supramolecular systems, the chirogenic processes of which have been comprehensively reviewed in different thematic publications.[2a,b,4a,6,7] To get a clear view of the respective supramolecular systems, readers are encouraged to refer to these review articles, books, and special issues. Irrespective of those scientific reports available in literature, the present book chapter deals with chirogenesis perspectives in supramolecular systems with its important applications.

This chapter presents our viewpoint on the major features of chirogenesis, rather than a comprehensive collection of literature. Therefore, we selected some representative examples including our research on the chirogenic processes occurring in several important areas, which can have and have already had an impact on advances and developments in innovative analytical methods and modern technologies. In part, focused on general methods for designing supramolecular chirogenic systems, we discussed its accounts/summary of exciting led applications in the viewpoint of chirogenesis. However, the currently booming research activities in this field provides new and exciting opportunities for designing unique chirogenic hybrid systems and developing new methods and approaches for the adaptive supramolecular chirogenic systems and related functional self-assembled materials in chiral sensory (sensing and recognition), transformations (asymmetric synthesis and catalysis), and their availability for future applications.

2. Design and Construction of Supramolecular Chirogenic Systems

Indeed, a better understanding of the perspectives of generating chirality in single molecules and multi-component systems and its related chirogenic processes will help researchers to find an optimal way to design a particular supramolecular chirogenic system toward the desired functionality. Considering natural systems and biomolecules like peptides and nucleic acids, the corresponding self-assembly process *via* non-covalent interactions resulted in chiral supramolecular structures such as a Deoxyribonucleic acid (DNA) double helix, secondary α-helical structure of proteins, heme proteins, and various (metallo) enzymes.[8] Importantly, these natural systems are regarded as the most predominant examples and/or models for the design and construction of various supramolecular chirogenic systems.

In this respect, designing an artificial supramolecular chirogenic system is considered to be an art of science. Even though these sophisticated assemblies can mimic the efficacy of nature, the involved chirogenesis and efficiency can be varied depending upon the nature and type of the supramolecule, specificity and versatility of guest binding with the host, and the effectiveness of communicating supramolecular interactions between the corresponding components. In particular, the chiral information is transferred either from host to guest or *vice versa*, where the host or guest (or both) should be chiral. This process may include several functions: chirality induction, modulation, sensing/recognition, and discrimination, hence being often confused.[2a,b,9] It varies depending on the process involved or applicability, thus making it important to consider the individual chiralities of host and guest and their effect on the physical, chemical, and chiroptical properties of the supramolecular complex (see Section 2.4).[2a,b]

Many various strategies are proposed and reported, so far, to introduce chirality in a molecule or finite assembly/system.[2b,6b,10] Accordingly, chirality in supramolecules can be broadly classified into three major categories based on the origin of their chirality, which are (a) *Intrinsic or inherently chiral supramolecular systems*: chiral arrangement of the achiral motifs or a segment with the help of substituents, endowing the symmetry plane loss of the whole molecule/structure due to the bending/twisting/curvature, thus resulting in distinguishable shapes/faces, and rendering them as the inherently chiral, e.g., bucky bowls, (b) *Extrinsic or induced chiral supramolecular systems*: covalent attachment of additional chiral fragments onto the achiral motifs or externally conjugated chiral molecule *via* non-covalent interactions, (c) *Chirally amplified/ modulated supramolecular systems*: increasing the chirality hierarchy *via* self-assembly of the molecular components of lower chirality to nano-/macroscopic level of structured supramolecular systems, and/or attained under the influence of some external (physical and/or chemical) factors.

Based on this classification, various methods were used to design and construct chiral supramolecular systems, having an inherent chiral domain like pockets/cavities/helix to be utilized in many functional applications related to chirality sensing, memory, and transfer.

2.1. *Inherently chiral supramolecular systems*

Inherent chirality is one of the specific descriptions of asymmetric property, which arises from the de-symmetrization of planarity in a molecule or a group of molecules, the generated/formed bend/twist/curvature, which clearly distinguishes the shapes/faces of the object in three-dimensional space. Inherent chirality can also be generated by non-covalent interactions like a hydrogen bonding. The curvature can also undergo racemization *via* the structural inversion, with the stability being dependent upon its inversion barrier and able to be separable/isolated ($\Delta G \leq 10$ kcal/mol) or only detected ($\Delta G \geq 10$ kcal/mol), for example by NMR.[2c,11] The inherent molecular chirality is described using various terms in chemical literature such as, bowl chirality (used in fullerenes and its fragments, corannulenes, and sumanenes), intrinsic chirality (carbon nanotubes), helicity (chirality of helical, propeller, or screw-shaped molecular entity), cyclo-chirality (directional array of chiral building blocks in cyclopeptides), and residual enantiomers (sterically hindered molecular propellers).[12a] Many biological molecules and their related process are inherently chiral in nature. For example, isomerization of retinal in vision process, where retinal is a photoreactive chromophore in rhodopsin that contains a protein and a achiral polyene chain. Upon light hitting the rod cell of eyes, the chromophore 11-*cis*-retinal (non-planar, inherently chiral) isomerizes to all-*trans*-retinal (planar, achiral) that makes changes in the attached opsin proteins' conformation and shape that leads to the generation of a nerve impulse. This isomerization occurs in a few picoseconds (10^{-12} s) or less. The conformational study about retinal shows that, although the retinal polyene chain contains no chirality centers, substantial effects or changes were observed in the Ultraviolet–visible (UV-Vis) and circular dichroism (CD) spectrum, in which the observations indicated that retinal exists as a racemic pair of two enantiomers in solution due to the presence of steric methyl groups, hence completely precluding its planar conformation to the curvature structure (inherently chiral) of its achiral polyene chain. Theoretical studies (*ab initio*) indicate that the inherent chiral conformation of 11-*cis*-retinal is the dominant factor in CD spectrum.[12b] Further, X-ray studies supported that the protein binding site accommodates an inherently chiral conformer of the chromophore (11-*cis*-retinal).[2c,12b]

Various classes of supramolecules can be inherently chiral depending upon their structural features, and these classes include salphen complexes, porphyrinoids (chlorine, phthalocyanines, and subphthalocyanines), cyclic amides, derivatives of sumanene, and corannulene, helicenes, and cavitands *viz.*, cyclodextrins (CDXs), calixarenes, pillararenes (PAs), hemicucurbiturils (HCs), cyclotriveratrylenes, resorcinarenes, and others (Figure 2).

2.2. *Chirally induced supramolecular (self-assembled) systems*

Generally, in supramolecular chemistry, induced chirality is one of the most important interdisciplinary divisions, where asymmetric information of a chiral host is transferred to an achiral guest (or *vice versa*) *via* non-covalent interactions (which is also called as chirogenesis). The host can

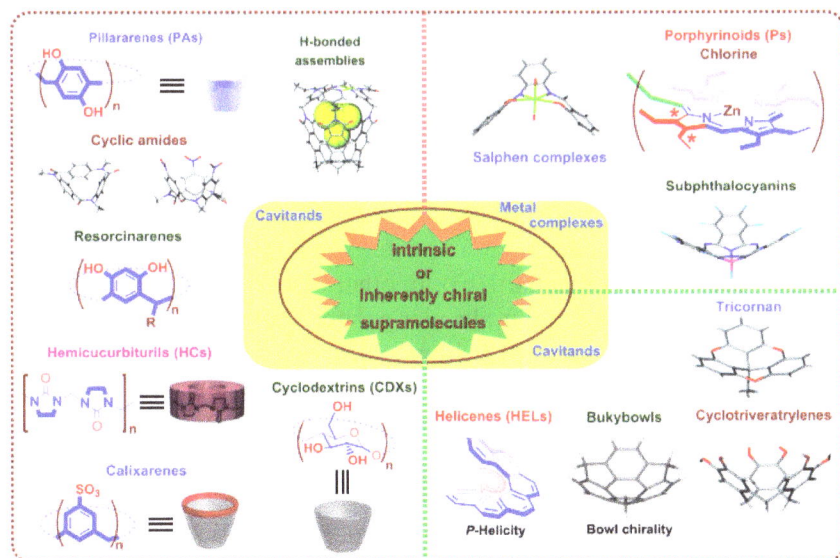

Figure 2. Examples of intrinsic or inherently chiral supramolecules.

be chiral macrocyclic compounds (CDXs, PAs, calixarenes, etc.), chiral pockets (enzymes, DNA, proteins, etc.), chiral cavity, and chiral nanostructures (self-assemblies, organogels, soft materials, etc.). In these supramolecular systems, chirality manifests in different forms and diverse range of hierarchical scales from molecular, macromolecular, and supramolecular levels to nano-/macroscopic level of structured self-assembled systems. In the formation of chiral assemblies or aggregates, the non-covalent interactions such as hydrogen bonding, van der Waals interactions, π–π stacking, hydrophobic binding, and so on are able to control the spatial arrangement of components in the assemblies, thus propagating the chiral information through the involved specific interactions and amplifying to diverse scale.[2b] This process is generally called chirality transfer. Since supramolecular chirality, as like natural, various artificial, and biomimetic systems, is very sensitive with respect to temperature, solvent, pH, photo-irradiation, ultrasound, and other external stimuli, those can also be effectively used to control the properties of supramolecular materials in a dynamic and reversible manner.[2b,10c]

For example, upon designing and constructing the self-assembly/gelation-induced supramolecular chirogenic systems, the corresponding functional groups and involved non-covalent interactions are efficiently transferred and amplified by molecular chirality to the supramolecular level in three different ways (Scheme 1). In the first case, only chirally pure small gelating molecular components were packed accordingly through the non-covalent interactions resulting in the chirality transfer and formation of homo-chiral self-assembled supramolecules (Scheme 1(a)). In the binary system, the molecular chiral gelator components are combined and co-assembled with an achiral dopant, also generating the chiral information transfer and homo-chiral supramolecular self-assemblies (Scheme 1(b)). In other cases, the symmetry breaking occurred in pure achiral molecular gelator components, yielding the corresponding self-assembled homochiral supramolecular systems (Scheme 1(c)).[2b,10c,13]

These supramolecular systems and functional soft materials of nano-/micro-scale obtained through the self-assembly process attracted much attention from the scientific community owing to prospective chirogenic applications such as chiral recognition, sensing, discrimination, and asymmetric catalysis. In particular, upon the self-assembly process, monomeric enantiopure molecules are able to be expressed in different ways of chirogenesis to form the various functional materials of diverse scale.

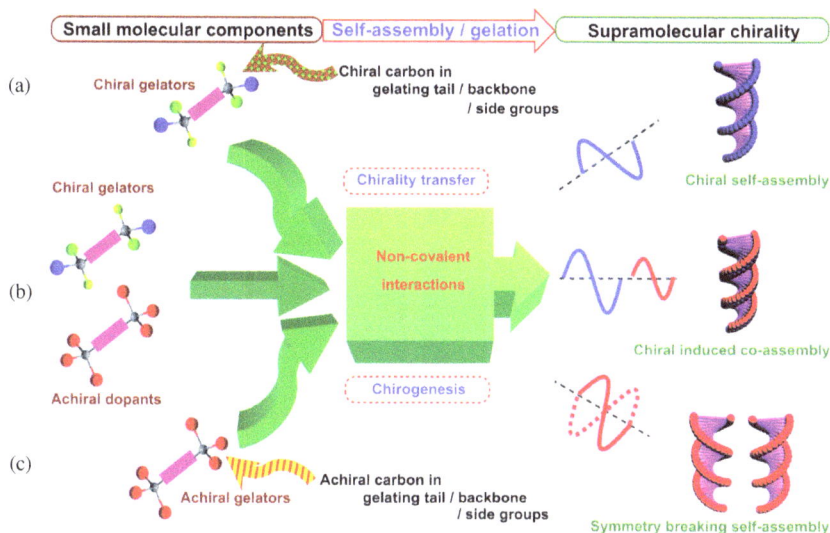

Scheme 1. General methods for the generation of induced chiral supramolecular chirogenic systems through the self-assembly process. Supramolecular chirality of self-assembled systems generated composed of (a) a chiral small molecular gelator components only, (b) chiral gelator and achiral dopants (binary system), and (c) achiral small molecular gelator components only.

Recently, a number of scientific reports and reviews have been published on the preparation and application of a wide range of self-assembled systems and supramolecular chirality.[2b,10c]

2.3. *Chirally amplified (modulated) supramolecular systems*

Amplification or modulation of chirality is one of the highly profound phenomena that was mainly utilized in molecular reactions (asymmetric and auto-catalysis)[2a,14] and self-assembling systems.[15] In the self-assembling systems, the chirality amplification process was defined in two basic concepts, which are the "sergeants-and-soldiers" principle and "majority" rule (Scheme 2).[16] These two principles can be clearly described as follows: the local chirality of a small enantiopure fraction decides the chiral sense of the entire self-assemblies. Mainly, the involved non-covalent interactions occurring during the self-assembling process lead to the

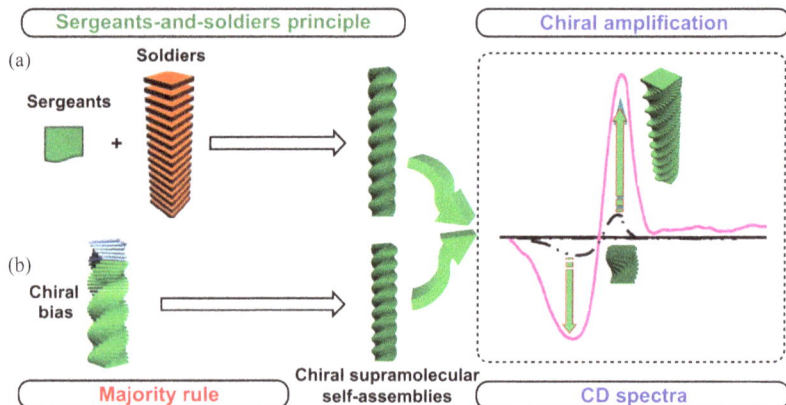

Scheme 2.   Illustration of chiral amplification/modulation process utilized to prepare the chiral supramolecular self-assemblies through (a) "sergeants-and-soldiers" principle, and (b) "majority" rules. Reprinted with permission from Ref. [2b]. Copyright 2018, Royal Society of Chemistry.

chirality transfer and amplification to form 1D/2D supramolecular assemblies/helical polymers/networks resulting in the manifestation of CD signals.

The utility of "sergeants-and-soldiers" principle in supramolecular self-assembly is based on the effect in which a few small chiral elements defined as "sergeants" control a large number of achiral components called as "soldiers" to form the whole chiral supramolecular system (Scheme 2(a)). In the case of "majority" rule, a slight initial excess of chiral bias leads to the transfer of its helical chiral sense to the entire self-assembling system, resulting in efficient amplification of chirality (Scheme 2(b)). These two elegant basic concepts of chiral enhancement process were widely used to obtain the variety of chiral supramolecular structures with the following approaches: (i) chiral-amplification in chiral analog-induced self-assemblies,[17] (ii) chiral amplification in binary self-assembling systems,[18] (iii) chiral amplification in nanoscale self-assembled structures,[19a] (iv) unprecedented chiral amplification in race-mate self-assemblies,[19b] and (v) chiral memory in supramolecular self-assembled polymers.[20] So far, these approaches have been compre-hensively analyzed and discussed in several reviews,[2a,b,10c,21] while here only selected supramolecular self-assembled systems are discussed, which are related to our subject.

2.4. *Overview of major applications*

Generally, supramolecular systems are used in various chirogenic applications, which are varied depending on the involved host–guest components (chiral/achiral) and the nature of the process occurring (Scheme 3).[2a,b,22] Concerning the chiral induction process, either a host or guest component should be chiral, with the optically active supramolecular complex formed upon the host–guest complexation. Meanwhile, steric factors and intermolecular interactions play an important role in the chirality transfer process, resulting in asymmetric conformational changes and subsequent chirality induction.

If both the host and guest components are chiral, the chirogenesis process is generally known as chiral recognition or discrimination. In this case, the complexation of a homochiral host with a pair of the enantiomeric guests generated two diastereomeric complexes, where these guests experience different intermolecular interactions resulting in distinguishable physico-chemical and spectroscopic outcomes. Using this process, the enantiomeric pairs can be recognized or discriminated easily with the conventional spectroscopic methods.[2a,22]

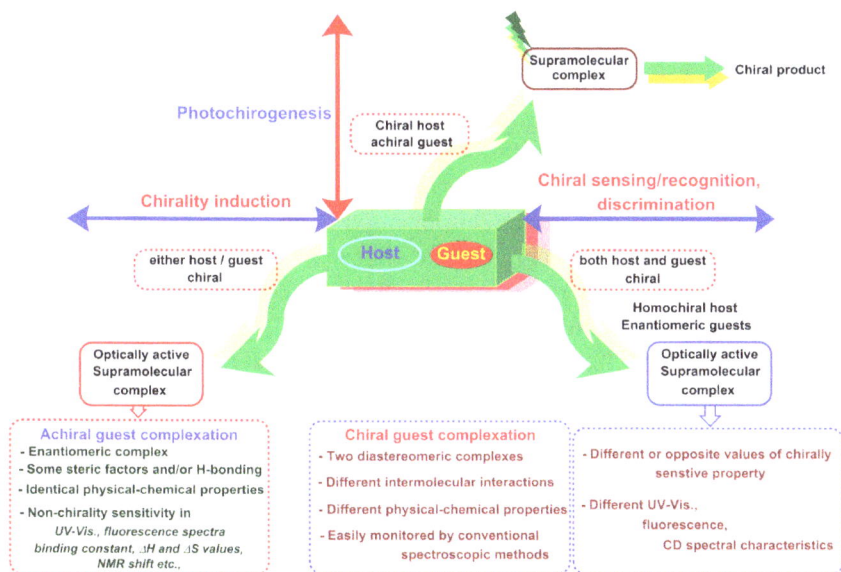

Scheme 3. Introduction to the processes involved in the major application of supramolecular chirogenic systems.

Generally, obtaining the enantiopure photoproducts is an arduous task in conventional photochemistry. This is because the reaction occurred in the electronically excited state characterized by its high reactivity and short life time, hence hindering the control of stereochemical reaction pathway. However, if the photoreaction is carried out in a supramolecular chirogenic host through the corresponding complexation *via* non-covalent interactions, energetic differentiation of the diastereomeric transition states and simultaneous chirality transfer in the excited state take place, leading to photoproducts with high enantiomeric excess.[6b] In Section 3, applications involving these chirogenic process by using different supramolecular systems are discussed.

Apart from the above-mentioned chirogenic applications, other types of supramolecular systems further involved in asymmetric catalysis (as chiral catalyst/additive/promoter, homogeneous, and heterogeneous) and functional materials (thin-films, metal surface, and soft-materials, etc.), optics and electronics, CPL-induced chiral assemblies, and biological related applications are briefly analyzed.[15c,e,21a,23]

3. Major Applications of Supramolecular Chirogenic Systems

Indeed, most naturally occurring biological structures are of specific handedness (DNA double helix, Tobacco mosaic virus, various metalloenzymes, etc.), known as homochirality, and its related biochemical processes are inherently chiral, being highly stereo-sensitive in life processes. These natural systems are the fascinating examples for researchers to design a system that can mysteriously translate/choose the homo-molecular chirality *via* tuning the nature of the specific supramolecular secondary interactions at various structural levels of molecules and organisms.[24]

In this respect, chirogenesis, being a process that transfers chirality to a molecule or whole system and *vice versa via* non-covalent interactions, is an effective tool to accomplish this task of chirality translation. Supramolecular systems are the most imperative designed assemblies found in nature (like enzymes) that can modulate many supramolecular chirogenic processes in various fields with the desired chirogenesis effect. The efficiency of this function can be easily tuned and optimized through the involved non-covalent interactions by means of external factors and stimulus (i.e., solvent, temperature, light, pH, additives, etc.).[2a,6a]

So far, a large number of various supramolecular chirogenic systems have been reported and applied in different practical applications, such as asymmetric catalysis, molecular and chiral discrimination, absolute configuration assignment, molecular devices, non-linear optics, polymer and materials science, molecular recognition phenomena at the nanoscale, and design of intelligent and responsive materials.[4a,6a]

In this chapter, we address and discuss several selected examples of supramolecular chirogenic systems and their chirality transfer properties applied in most emergent scientific fields, such as chiral recognition, sensing, and photochirogenic reactions. Current developments, challenges, and prospects are highlighted in this topic as well.

3.1. *Supramolecular chirogenic sensing and recognition*

Natural systems are fascinating models to endeavor us to create new methods and strategies for the synthesis of enantiomerically enriched compounds and develop highly selective recognition systems to discriminate and quantify its chirality. These enantiopure compounds found many unprecedented applications in fine chemical, flavor, pharmaceutical, and fragrance industries.[6a,25] In general, chiral sensing and recognition of small molecules using supramolecular systems have more advantages in comparison to conventional methods (chiral High-performance liquid chromatography (HPLC), Nuclear magnetic resonance NMR), X-ray crystallography), becoming an attractive and challenging research target in recent years.

In molecular recognition chemistry, the involved interaction and chemistry associated with a host molecule recognizing a partner guest molecule produce different spectral outcomes. If a host molecule is achiral, the subsequent interaction with a chiral guest molecule (and *vice versa*) resulted in chirality transfer through non-covalent interactions (chirogenesis), producing enantiomeric supramolecular complexes, that possess identical physio-chemical characteristics and opposite sign values of chirally sensitive properties. The generated supramolecular chirality through asymmetry transfer can be detected by CPL, CD characteristics, and specific optical rotations. This phenomenon is mainly used to observe and quantify chirality of a host–guest molecule.

Of the vast number of host–guest and self-assembly systems, some of them and its hybrid conjugates are especially useful to be employed as a supramolecular host for the chirogenic applications as a consequence of them having specific and tunable non-covalent interactions. Some representative systems possessing especially notable chirogenic properties with in-depth investigations are discussed in the forthcoming subchapters.

3.1.1. *CDX-based systems*

CDXs are one of the foremost studied systems among the supramolecular family. CDXs are naturally occurring cyclic oligosaccharides composed of 6–8 α-*D*-glucose units (6 units for α-CDX, 7 units for β-CDX, and 8 units for γ-CDX) connected in a cyclic manner (through the α-1,4-glycosidic linkages) to form truncated cone-shaped hydrophobic cavities with different sizes (top/bottom diameters of the cavity: 4.7/5.3, 6.0/6.5, and 7.5/8.3 Å for α-, β-, and γ-CDX, respectively), which can accommodate a wide range of the organic guests of different bulkiness in the cavity in an aqueous solution. The complexation is mainly driven by hydrophobic and other supramolecular interactions (hydrogen bonding, Van der Waals, Coulombic, and π–π stacking binding, etc.). Generally, these CDXs are water-soluble, inherently chiral, spectrally transparent (over UV–Vis region) hydrophobic cavities and can be easily functionalized on both sides with various chemical groups. As a consequence of these advantages, the CDXs are extensively used as a unique supramolecular host in the various chirogenic applications, being mainly employed as chiral selectors and in drug delivery systems.[2a,6b,26]

Native CDXs tend to form inclusion complexes with hydrophobic guests through relatively weak and non-specific hydrophobic interactions, hence transferring the chiral information to the guest inefficiently.[27] The poor chirogenesis and chiral selectivity of native CDXs with a chiral guest have been evaluated by the determination of complex stability constants and thermodynamic parameters such as enthalpy and entropy values for the inclusion complexation of both enantiomers.[28]

Generally, the inherently chiral cavities of native CDXs provided a low to moderate chirogenic ability toward the complexed guests. This serious drawback can be properly addressed upon selective/per-modification of the 1° or 2° hydroxyl groups in the narrow or wider rim of CDXs with appropriate functional groups/chromophores (Figure 3).[26a,29] Specific and selective modification of CDXs with chromophores results in enhancement of the chiroptical properties, making them effective chiral auxiliaries and supramolecular chirogenic hosts to be used for selective sensing and recognition of a wide variety of chiral molecules.[30]

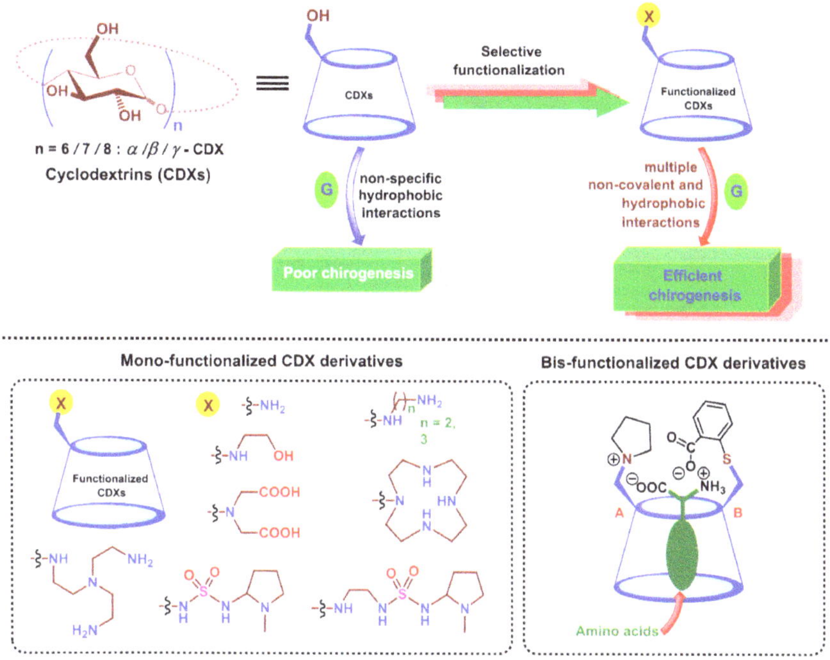

Figure 3.   Structure of native CDs and the selectively modified CD derivatives used as chirogenic host for chiral sensing/recognition and discrimination of chiral molecules.

Thus, the functionalized CDXs and their metal complexes have been successfully employed as supramolecular chiral sensing and recognition receptors for many enantiopure amino acids (Figure 4).[26a] Enhanced enantioselectivity of these CDX derivatives in comparison to unmodified CDXs is mainly due to their ability to interact with the included enantiomers in the chiral hydrophobic cavity differently; thus, the formed complexes have a different chirogenic effect, which can be monitored by different spectroscopic techniques like UV–Vis absorbance, fluorescence, CD spectroscopy, NMR, etc.

Recently, Yang and co-workers prepared γ-CDX-CB[6]-co-wheeled [4]pseudorotaxanes, **Rot-bp-axa/bp-axb/np-ax** *via* the self-assembly-based rotaxanation strategy and achieved enhanced chiral discrimination in comparison to native γ-CDXs by the factor of several hundreds (Scheme 4).[31] This rotaxanes contains axels, **bp-axa/axb and np-ax**, which were designed based on "three-point" complexation model containing an achiral hydrophobic part (biphenyl or naphthyl rings) and charged

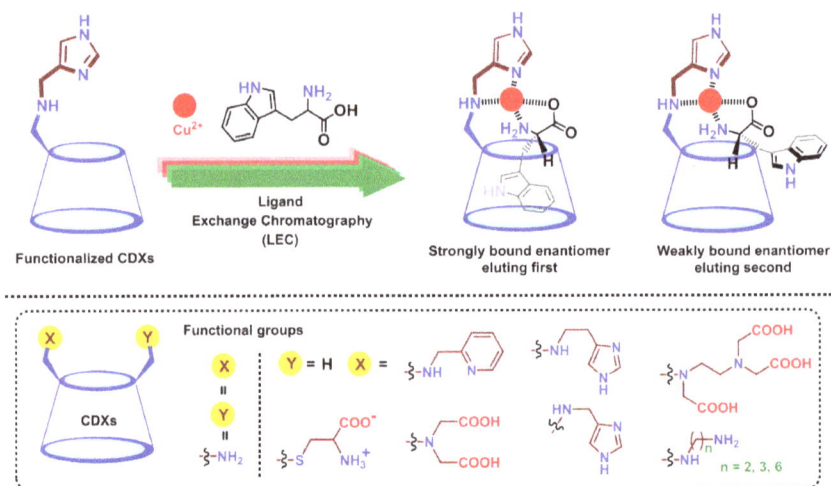

Figure 4.   Metal complexes of selectively modified CD derivatives used as chirogenic host for chiral sensing/recognition and discrimination of chiral molecules.

tethers (tetramethyldiamines) on both sides, thus forming γ-CDX-CB[6]-co-wheeled [4]pseudorotaxanes spontaneously upon self-assembly with γ-CDX and CB[6]. The implanted achiral aromatic axel in the hydrophobic cavity of CDs provided a highly unsymmetrical chiral binding site and additional π–π and ion–dipole interaction for chiral amines with remarkable improvement in the chiral discrimination. Naphthalene co-wheeled [4]pseudorotaxane, **Rot-np-ax3**, showed a high binding affinity toward α-methyl-1-naphthalenemethylamine, **NMA**, with the chiral discrimination factor ($K_{R(S)}/K_{S(R)}$) being up to 5.05. Similarly, the relatively weak and non-directional hydrophobic, chirogenesis sensing ability of β-CDs upon interaction with underivatized proteinogenic amino acids and the corresponding enhancement upon modification of a helically chiral azonia[6] helicene chromophore have been reported (see the detailed discussion in the Section 3.1.3).[32]

3.1.2. *Porphyrinoid-based systems*

The family of macrocyclic tetrapyrrole analogs (i.e., porphyrinoids) is quite large, which includes porphyrins (Ps), chlorins, phthalocyanines, and corroles (contracted porphyrins). These macrocycles are ubiquitous and invaluable molecular species in living and artificial systems.

(a) Axels

Bp-axb : n = 6
Bp-axa : n = 4

Np-ax

(b) Rotaxanes

γ-CDX

π-π interaction

CB[6]

CB[6]

Ion-dipole interaction

Rot-bp-axa/axb   Rot-np-ax

(c)

NMA

(d)

θ / mdeg

Wavelength / nm

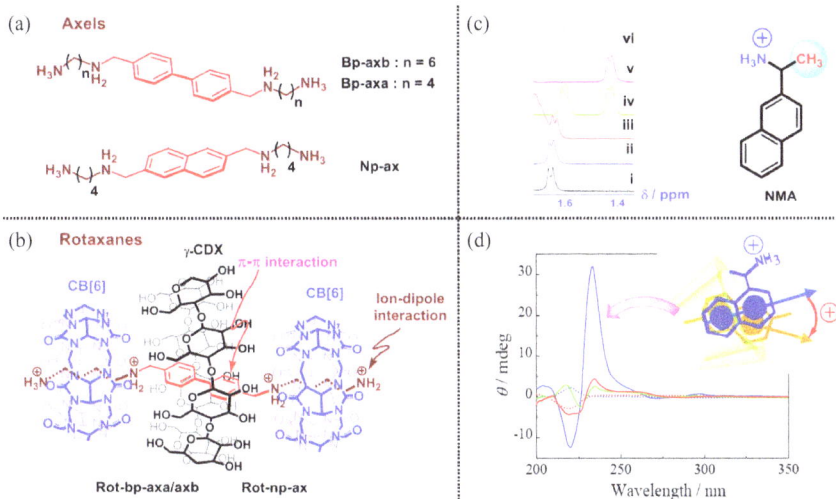

Scheme 4. Structure of (a) axels, **bp-axa/-axb and Np-ax**, (b) illustrated hybrid system of γ-CDX-CB[6]-co-wheeled [4]pseudorotaxanes, **Rot-bp-axa/bp-axb/np-ax**, (c) partial ^{1}H NMR spectra of D_2O solution containing (i) 10 mM racemic **NMA**, (ii) 10 mM racemic **NMA** and 10 mM γ-CDX, (iii) 10 mM racemic **NMA**, 10 mM **bp-axb**, and 10 mM γ-CDX, (iv) 10 mM racemic **NMA** and 10 mM **Rot-bp-axb**, (v) 10 mM **(R)-NMA** and 10 mM **Rot-bp-axb**, (vi) 10 mM **(S)-NMA** and 10 mM **Rot-bp-axb**, and (d) enantioselective recognition process of **Rot-np-ax** with **NMA**. Reprinted with permission from Ref. [31]. Copyright 2018, Royal Society of Chemistry.

The porphyrinoid structure is an extraordinarily stable system containing four pyrrole rings, which are interconnected at the α carbon atoms through the methine units as the corresponding linkages. Four nitrogen atoms occupy the interior part of central cavity that can easily be co-ordinated to metal ions (Fe, Co, and Mg prevail in biological systems) in a tetradentate fashion. Porphyrinoids are efficient chromophores absorbing light in the UV–Vis region of the electromagnetic spectrum and appearing as deeply colored (ranging from red to violet) compounds.[33]

In general, the P analogs are expedient chromophores for chirogenic processes in the host–guest system as well, thus having a unique and/or tunable spectral and physicochemical properties like a wide window of light absorption/emission and high quantum yield. Besides, additional advantages of the P structures include easy handling, metal coordination, feasible chemical modification in the exterior core, great biological importance, selective binding with various analytes, and wide applicability.[34] These Ps are extensively involved in a large number of biological processes in living organisms such as respiratory gas (oxygen) transport (hemoglobin), photoreaction centers (chlorophylls) in photosynthesis, biomolecular redox catalysts (cytochromes), vitamin B_{12}, etc. As consequences of these tunable and multi-functional properties and other advantages, porphyrins are widely used in many artificial/synthetic biomimetic model systems, such as light-harvesting antennae, energy/electron donors and acceptors, catalysts, selective coordination and recognition species (cations, anions, and neutral analogs), and supramolecular chirogenic systems (chiral selectors/ differentiator and asymmetric catalysis).[7b,35]

The metal-free porphyrin ligands (free base) are able to form square planar coordination complexes with metal ions, which are naturally achiral p, thus having D_{2h} or D_2 and D_{4h} or D_4 symmetries. Introduction of chiral substituents or tethering with intrinsically chiral structures in the ligand periphery (either at the *meso* position or β-/γ-carbon of pyrrole ring) can de-symmetrize the existing planarity and the P analogs can attain inherent chirality.[7b,36] However, in the case of monomeric chiral porphyrins, the corresponding optical activity is relatively small (or even zero), hence leading to limited investigations and applications.[36b,37]

To overcome this problem, it was found that bis-porphyrin (**bis-P**) systems can be effectively used. In these systems, two Ps are covalently or non-covalently connected through a chiral/achiral linker resulting in a closer distance between two chromophores and, as a consequence, coupling of the corresponding electric dipole transition moments. This leads to splitting the excited state energies to an extent and shows an exciton coupling CD signal with a notable amplitude. In general, exciton coupling theory is used to rationalize the CD spectral properties of **bis-P**s, like the signal intensity, which is related to the spatial arrangement and corresponding conformations of porphyrins as in covalently attached and in non-covalently self-assembled systems.[38] Further, suitably derivatized **bis-P**s (Figure 5) and self-assembled chromophore systems have been developed to an extent to be mainly used as supramolecular chirogenesis hosts for chiral sensing and recognition, asymmetric catalysis, and conformational research on bio-polymers.[2a,7b,39]

As **bis-P** systems have a unique structural advantage in terms of versatile synthetic modifications and spectral and physicochemical characteristics, Inoue, Borovkov and co-workers designed achiral zinc and magnesium complexes of **bis-P**, **bis-P(Zn)/(Mg)** (Scheme 5) containing a

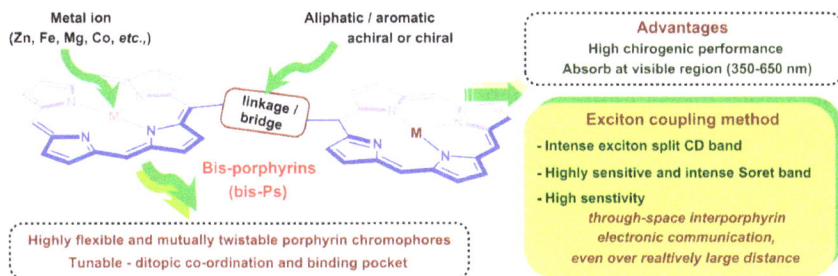

Figure 5. Advantages of bis-porphyrin (**bis-P**) system in chiral sensing and recognition.

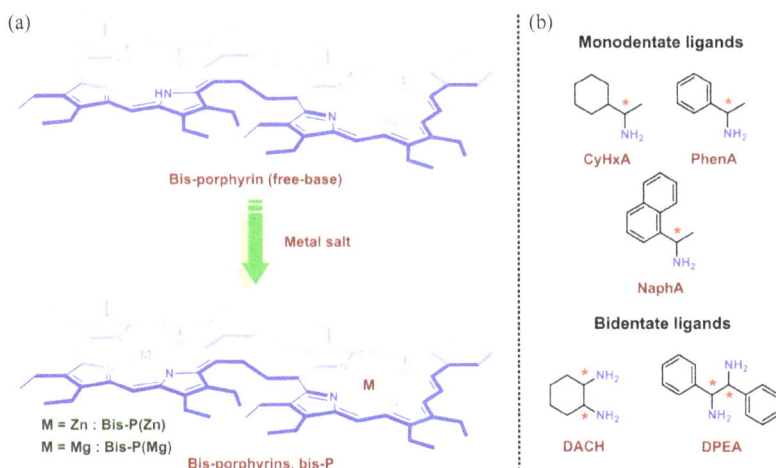

Scheme 5. Structure of (a) flexible ethane bridged bis-porphyrin (free base) and its metal complexes, **bis-P(Zn)/(Mg)**, and (b) selected chiral guests utilized for sensing and recognition.

flexible ethane bridge and systematically studied their chiral sensing properties, effect of solvent, temperature, structure of the chiral guest (size, number and nature of co-ordinating group), and its stoichiometry (of mono- and bi-dentate chiral guest) based on supramolecular chirality induction and inversion.[40–42] These supramolecular hosts, **bis-P(Zn)/(Mg)**, have been used to effectively sense the guest's chirality and assign their absolute configuration of enantiomerically pure amines, diamines, alcohols, amino alcohols, and C-protected amino acids as in solution and in solid state. Further, the authors clearly demonstrated the mode of binding of chiral guests (mono- and bidentate) to **bis-P(Zn)/(Mg)** for supramolecular chirality induction and inversion as supported by the thermodynamic parameters and the CD, NMR spectroscopic analysis.[40–42]

Indeed, less planar bis-porphyrin (free base) can easily generate a highly planar **bis-P(Zn)/(Mg)** complex, which has a strong $\pi-\pi$ interaction between the highly planar chromophoric moieties, and thus exists as a stable *syn* face-to-face conformation (achiral) in non-polar and noncoordinating solvents even at high temperature (up to 110°C).[40a] This enthalpically stable *syn* form was changed to the extended *anti* conformation upon addition of suitable coordinating ligands (mono-/bidentate, chiral/achiral) with observable spectral characteristics in UV–Vis and CD spectroscopy (Figures 6(a) and (b)).[40–42] Upon addition of chiral guests to **bis-P(Zn)**, the binding takes place at the Zn central ion of achiral *syn* form resulting in the de-symmetrization process *via* destroying the existing strong $\pi-\pi$ interaction between the porphyrin subunits to form the chiral *anti* form with the effective chirogenesis (chirality transfer) occurring from a chiral guest. The effective complexation, conformation change, and chirogenic performance were clearly observed at the Soret absorption band of **bis-P(Zn)**, with the hyperchromic as well as bathochromically shifted well-resolved split B electronic transition in UV–Vis spectra and bisignate (oppositely signed) Cotton effects (CEs) in CD spectra upon coordination with the chiral guests (Figures 6(b) and (d)).[40a–e,41,42a,42c–e]

Induced supramolecular chirality in **bis-P(Zn)** was mainly achieved by the gradual conformational change from the achiral *syn* form to the chiral *anti* form upon supramolecular complexation with enantiopure guests. Further, the mechanism of chirality induction was based upon formation of the right- or left-handed screw structures, which were successfully applied for assigning the absolute configuration of various chiral compounds (Figure 6).[40a–e,41,42a,42c–e]

The formation of optically active supramolecular complexes was strongly dependent on various external factors and nature of the enantiopure guests. As expected, the formed chiral supramolecular screw structures (in *anti* form) are more stable at lower temperatures, hence achieving effective chiral information transfer from the enantiopure guests.[40a,42c] The absolute configuration of the enantiopure guest plays an essential role in the process of chirality induction and the chirogenesis related to the screw formation in **bis-P(Zn)**, where (*S*)-enantiomers induced positive chirality

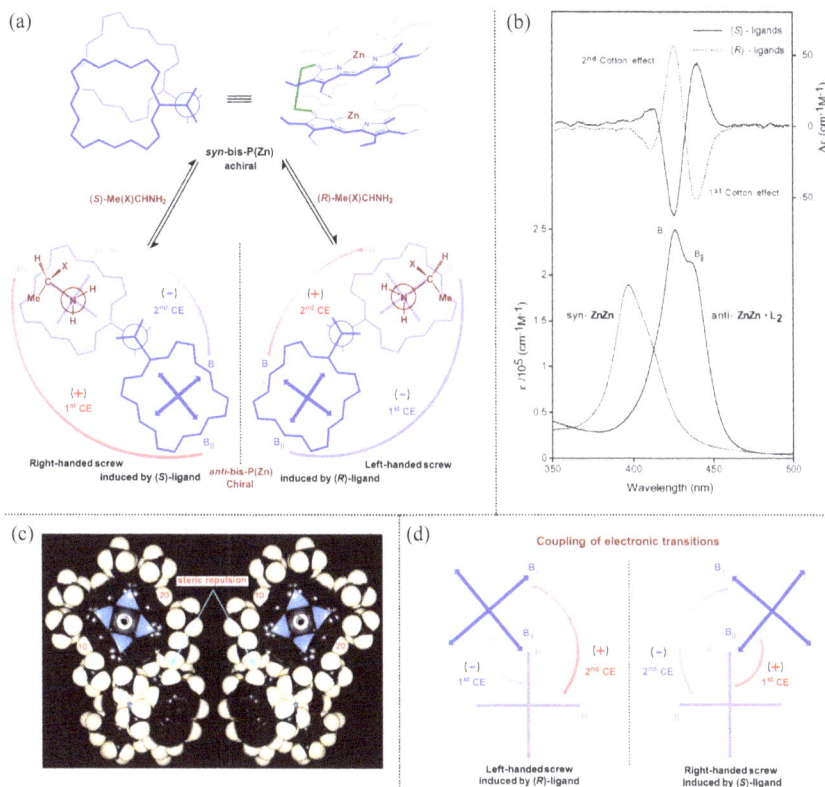

Figure 6.   (a) Schematic illustration of induced supramolecular chirality in **bis-P(Zn)**, achiral *syn* form co-ordination with enantiopure guests to form chiral *anti* form, (b) respective UV–Vis and CD spectral changes, (c) space-filling model (CPK model) of right- or left-handed screw structures (the ligand shown is 1-(1-naphthyl)ethylamine), and (d) orientation of the Soret band electronic transitions in chiral complexes and its coupling electronic transitions of supramolecular chirality induction. Reprinted with permission from Ref. [42e]. Copyright 2004, American Chemical Society.

(sign of the first CE) since a right-handed screw is formed and *vice versa* with (*R*)-enantiomers (induced negative chirality due to a left-handed screw) (Figures 6(a) and (b)).[40a–e,41,42a,42c–e] The CD amplitude was also dependent on the structure and nature of functional group involved in the coordination event. Generally, amines have higher binding affinity than alcohols, in which primary amines bind strongly than secondary amines. Similarly, chiral guests with aromatic or extended rings interact strongly with **bis-P(Zn)**, and thus efficiently transfer the chiral information, resulting in enhanced CD signals. Further, the delivery of chirogenesis to **bis-P(Zn)** also depends on the distance of the interacting group with respect to the asymmetric carbon atom and steric factors.

Phase transition (solution–solid) and medium (solvent–solute interactions) are also an effective tool for controlling the chirogenesis in **bis-P(Zn)** to increase/decrease and inverse the supramolecular chirality by switching the chirality induction mechanism.[40e,41a,c,42d]

Interestingly, the stoichiometry of chiral guest with respect to **bis-P(Zn)** is another important factor to control the supramolecular chirality induction and inversion (Figure 7). For example, addition of chiral bidentate guest, (*R*,*R*)-DPEA to *syn*-**bis-P(Zn)** at the low molar excess region resulted in the formation of a 1:1 tweezer type supramolecular complex with the right-handed screw.

However, the complex helicity was switched upon increasing to the excess amount of (*R*,*R*)-DPEA, which formed the extended left-handed screw of 1:2 *anti* complex.[40c] These two chirogenic effects were easily observable by UV–Vis and CD spectra (Figures 7(b) and (c)) at the Soret band region and unambiguously rationalized by analyzing the electronic transition coupling in the corresponding 1:1 and 1:2 complexes. Whereas, in the case of monodentate chiral guests, **bis-P(Zn)** forms 1:1 (as an intermediate) and 1:2 supramolecular complexes with the well-defined

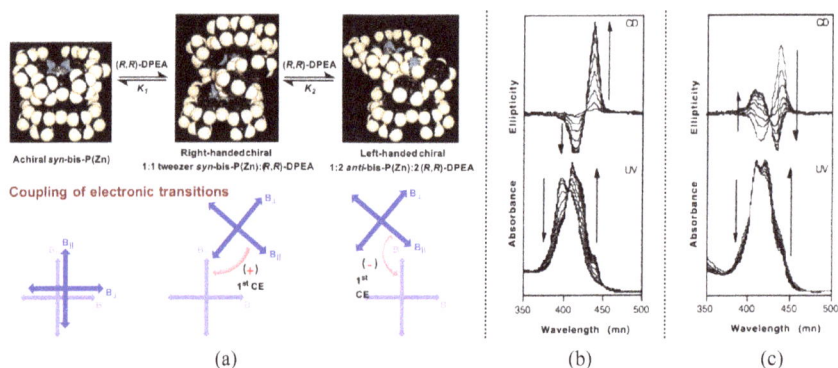

Figure 7. (a) CPK molecular models and its coupling electronic transitions of supramolecular chirality induction and inversion in achiral *syn*-**bis-P(Zn)** by (*R*,*R*)-DPEA. (b) CD and UV–Vis changes of **bis-P(Zn)** upon addition of (*R*,*R*)-DPEA at the low ratio from 1:0.12 to 1:16.5 and (c) high ratio from 1:112 to 1:4866 ligand molar excess regions. Reprinted with permission from Ref. [40c]. Copyright 2002, American Chemical Society.

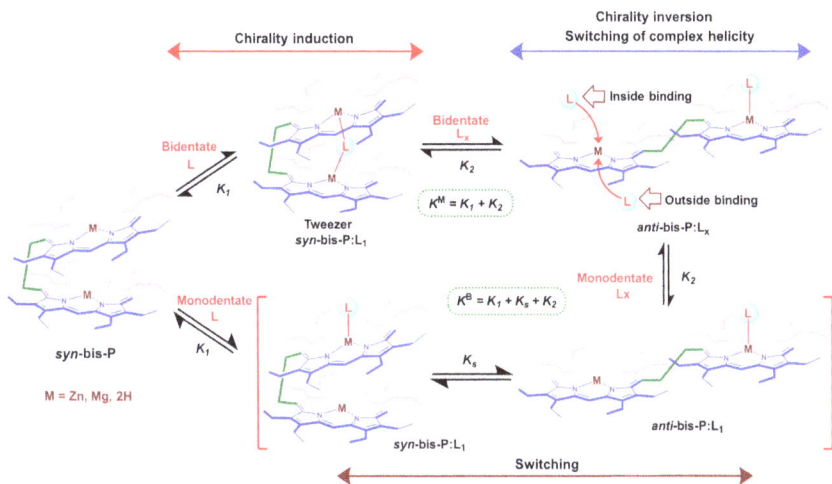

Scheme 6.   Schematic illustration of mono/bidentate chiral guest dependent 1:1 and 1:2 supramolecular complexation with **bis-P(Zn)**.

left-/right-handed screws, the supramolecular chirality was intensified but not inverted upon addition of the chiral guest excess (Scheme 6).[40c,d]

Further, a coordinating metal ion in **bis-P** also plays a significant role in chirogenic differentiation of chiral guests. Thus, **bis-P(Mg)** was directly used to assign the absolute configuration of chiral mono-alcohols at room temperature based on the CD exciton chirality method.[40f] Like **bis-P(Zn)**-monoamine systems, **bis-P(Mg)** binds chiral monoalcohols generating a bisignate CD signal in the Soret region. Also, (*S*)-enantiomers formed a right-handed screw structure, with the positive CD sign, whereas the opposite situation was observed with (*R*)-enantiomers. As the sequential utility of **bis-P(Zn)/(Mg)** for the stereochemical assignment of chiral guests (amines and alcohols), **bis-P** (free base) was also used as a supramolecular chirogenic host for the absolute configuration determination of enantiopure acids.[42b]

Interestingly, besides the porphyrin based systems these researchers designed and synthesized the intrinsically chiral bis-chlorin, **bis-C** with four stereogenic centers. Surprisingly, **bis-C** was resolved as only two enantiomers and their absolute configuration was assigned and high optical activity was rationalized through a combined spectral, crystallographic, and theoretical analysis (Figure 8).[41d,e,42a] The enantiopure **bis-C** was used as a unique two-point interaction supramolecular chiral host for

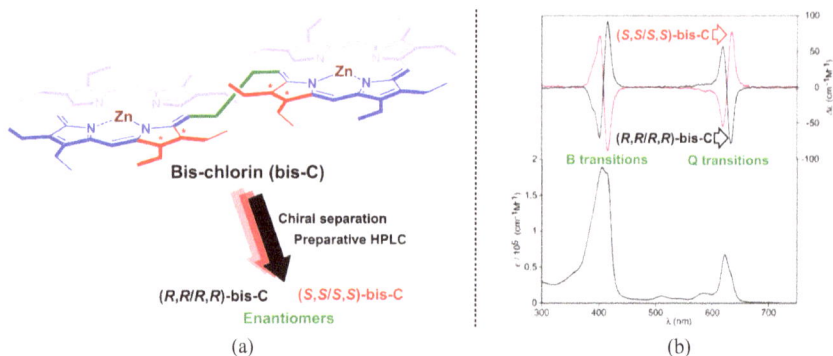

(a)

(b)

Figure 8. (a) Structure of bis-chlorin, **bis-C** and (b) experimental absorption (bottom) and CD (top) spectra of (*R,R/R,R*)-**bis-C** (first eluting, black line) and (*S,S/S,S*)-**bis-C** (second eluting, red line) in CH$_2$Cl$_2$ at room temperature. Reprinted with permission from Ref. [41e]. Copyright 2005, American Chemical Society.

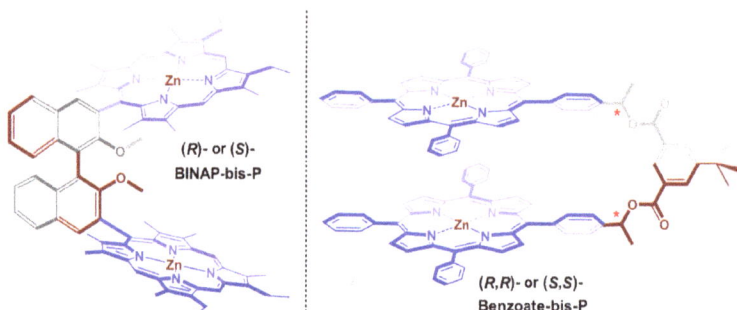

Scheme 7.   Structures of chiral **bis-P**s.

the antipodal amine guests, to investigate the host's conformational response, as well as chiral modulation and recognition.[41d,42a] The enantioselectivity was switched and controlled by varying the substituent's bulkiness at the stereogenic center.

Another structurally divergent **bis-Ps** was also reported with a chiral unit as the corresponding covalent linker (Scheme 7), which was used for chiral sensing and catalysis. Chiral 1,1'-binaphthyl and benzoate bridged **bis-P(Zn)s**, **BINAP-bis-P**,[43a] and **benzoate-bis-P**[43b] have been prepared and applied for chirality determination of diamines by the exciton coupling method. Chiral **bis-P**, **BINAP-bis-P** shows good affinity for achiral

α,ω-diamines, $H_2N(CH_2)nNH_2$ (n = 6, 8, 10, 12; K_a = 10^5–10^6 M^{-1} in CH_2Cl_2 at 15°C), resulting in the significant bisignate CD spectra (due to the exciton coupling between two **Ps**), and its intensity depends on the length of diamine,[43a] whereas chiral **bis-P**, **benzoate-bis-P** functioned as a highly sensitive chiral shift reagent for the determination of the enantiomeric purity of various chiral bidentate ligands (diamines, aziridine, and isoxazoline) at the microgram level using CD and NMR spectroscopies.[43b] Enantiopure **benzoate-bis-P** shows less intense bisignate CD signals at the Soret band region, while upon interaction with enantiopure diamines (i.e., cyclohexane-1,2-diamine (DACH)), the bisignate CD signals were considerably increased. As the enantiopure **benzoate-bis-P** complexed with respective chiral DACH through the induced-fit type of binding, it resulted in a rigid conformation for the two **P** rings and enhanced its electronic transition interactions. Then the sign sequences of the observed CD exciton couplets were used to determine the chirality of the diamines.

A new **bis-P** based supramolecular chirogenic host, **biphenol-bis-P** was synthesized with axially chiral 2,2′-biphenol as a bridge and was used as a point-to-axial chirality transfer hydrogen bonding receptor for various chiral amines (Scheme 8).[44] A rapid free rotation around the central single bond at the biphenol core resulted in both *P* or *M* helical conformers. However, tethered bulky P moieties imposed the corresponding steric hindrance leading to an energetically favored conformer. Hence, **biphenol-bis-P** complexed with certain enantiopure alkyl and aryl amines give intense exciton coupling CD corresponding to the Soret band of **Ps**. The specific helicities for the biphenol group predominate with respective chiral amines complexation. In the case of (*S*)-amine, a negative CD signal was observed, while (*R*)-amines produce positive CD signals. In spite of the differential responses of **biphenol-bis-P** with chiral amines, various steric factors play a major role in the formation of supramolecular complexes, and a conformational analysis is needed to rationalize the enantioselective binding of chiral guests with *M* and *P* helical conformers (Scheme 8).

Further, Rath and co-workers reported the flexible diphenyl ether bridge containing **bis-P** system, **DPE-bis-P**, which was specifically used as a substituent size-dependent chirogenic differentiation host for chiral diamines (Scheme 9).[45a] The same authors also reported the stoichiometry-dependent supramolecular chirality induction in the dibenzothiophene-bridged **bis-P** system, **DBTP-bis-P** upon interaction with chiral vicinal diols.[45b] The corresponding hydroxyl groups are able to coordinate with two Mg(II) center

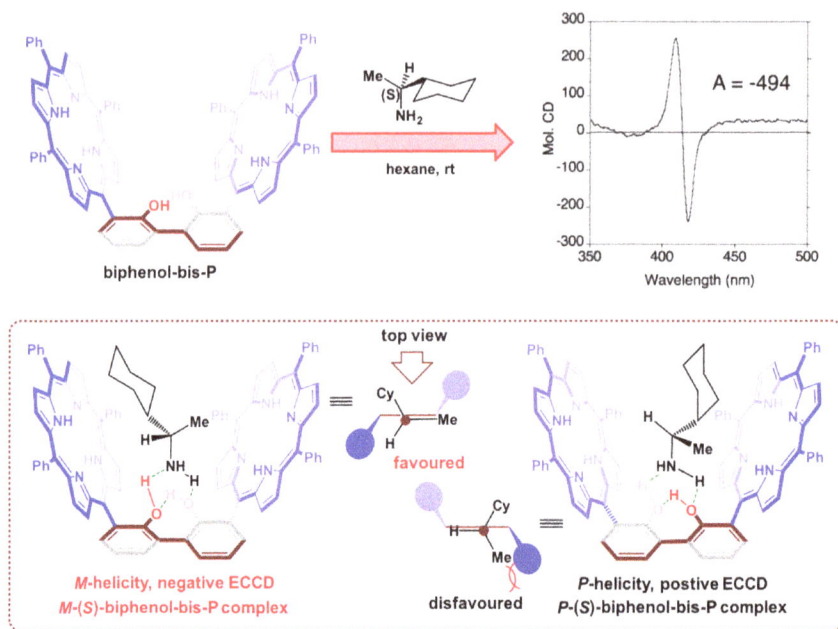

Scheme 8.   Proposed working model for assigning the absolute stereochemistry of chiral amines. Complexation of (*S*)-cyclohexylethylamine with **biphenol-bis-P** is illustrated for the host molecule with both *M* and *P* helicity. The *P*-(*S*) complex leads to the positioning of the medium CH_3 group in a sterically encumbered region compared to that in the *M*-(*S*) complex, which places the smallest group (H) in the same comparable location. The experimental results yielded a strong negative ECCD spectrum that corroborated the predicted assignment. Reprinted with permission from Ref. [44]. Copyright 2014, American Chemical Society.

of achiral **DBTP-bis-P** in the *endo–endo* manner through the inter-ligand hydrogen bonding interactions, to achieve the efficient chirogenesis from a chiral guest to an achiral host (Schemes 9(b)–(d)).

Unique and synergistic characteristics of **bis-P**-based host systems were significantly grown in recent years and contributed an important part in the supramolecular chiral sensing and recognition, where chirality induction was achieved by the chirogenesis process through the coordination (or other types of binding) of enantiopure guests. These **bis-Ps** are a rational model system for supramolecular chemists and spectroscopists to design an efficient system to determine the absolute configuration of chiral compounds and carry out its related future research in various real-time applications.[2a,7b,35b,39]

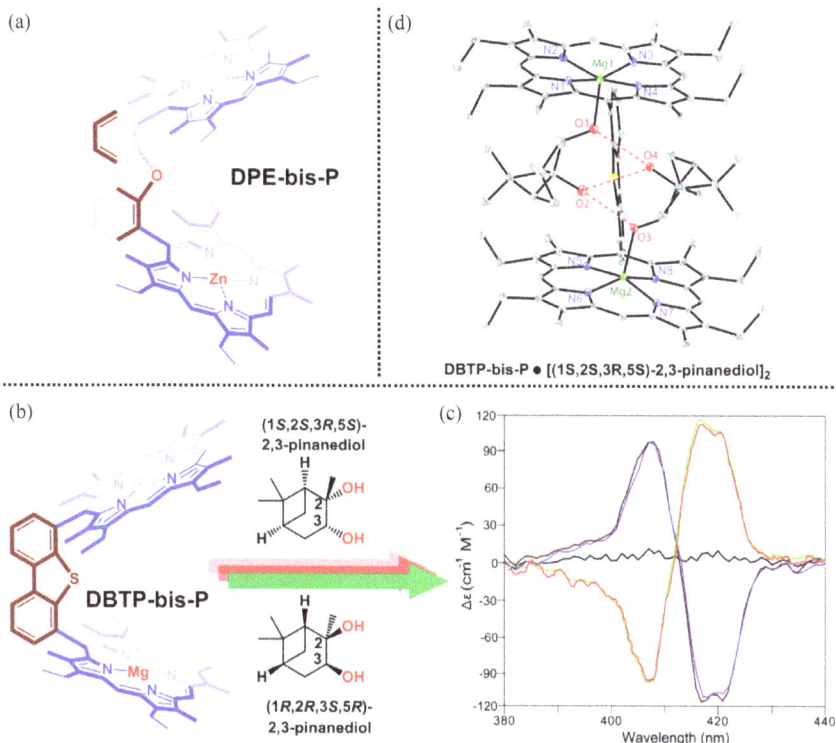

(a)

DPE-bis-P

(b)

DBTP-bis-P

(1S,2S,3R,5S)-2,3-pinanediol

(1R,2R,3S,5R)-2,3-pinanediol

(c)

(d)

DBTP-bis-P • [(1S,2S,3R,5S)-2,3-pinanediol]$_2$

Scheme 9. (a) and (b) Structures of diphenyl ether- and dibenzothiophene bridged bis-Ps, **DPE-bis-P**, and **DBTP-bis-P**, respectively, (c) calculated CD spectra of **DBTP-bis-P** (black), **DBTP-bis-P**·[(1S,2S,3R,5S)-2,3-pinanediol]$_2$ (brown), and **DBTP-bis-P**·[(1R,2R,3S,5R)-2,3-pinanediol]$_2$ (green) and correspondingly observed CD spectra blue line and red line, (d) a perspective view of **DBTP-bis-P**·[(1S,2S,3R,5S)-2,3-pinanediol]$_2$ showing 50% thermal contours for all non-hydrogen atoms at 100 K (H atoms have been omitted for clarity). Reprinted with permission from Ref. [45b]. Copyright 2014, Royal Society of Chemistry.

3.1.3. *Helicene-based systems*

Another prospective class of chirogenic supramolecular hosts, which attracted much attention from the scientific community is helicenoid structures. Helicenes (**HELs**) are polycyclic aromatic compounds in which benzene rings are ortho-fused or other aromatics are angularly annulated, thus forcing it to adapt a strained non-planar screw-shaped skeleton possessing helical chirality.[46] This helicity is dependent on the annulated ring size and hetero atom insertion (if any). For example, **HELs**

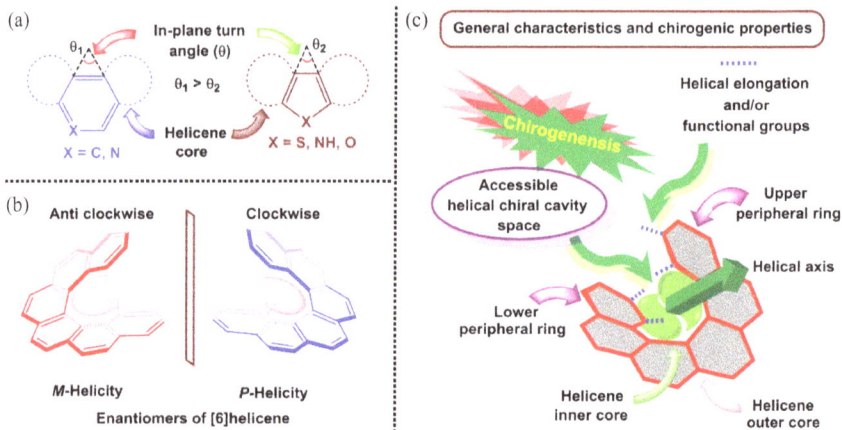

Figure 9.   Schematic representation of (a) in-plane turn angle, (b) helicity, and (c) enantiomers of [6]helicene, **[6]HEL**. General characteristics, chirogenic properties, explanation of terms used in helicene structures, and helical chiral cavity (shaded area).

containing all hexagonal (benzene and pyridine) aromatic rings, the number (n) of rings equal 6 and is essential to make them helically chiral with the in-plane turn angle of 60°, whereas the helicenes with at least one pentagonal (five-membered) heteroaromatic ring need n > 6 to become intrinsically chiral, with the in-plane turn angle being considerably smaller, of 45° (thiophene), 35° (pyrrole), and 32° (furan) (Figure 9). Helically arranged both terminal or peripheral rings influenced its steric hindrance toward the inner core of **HEL**. The suitably modified functional groups in the terminal or peripheral moieties and inner core helical chiral cavity effectively involve in various supramolecular chirogenic processes like chiral discrimination and "helical chirogenic templating" in asymmetric catalysis and organocatalysis. The nature of substituent group, structural features, and its position in a helicene molecule may affect the binding orientation of guest to host, and the involved steric influence of host to guest controlling their efficiency in chirogenic processes.[6f]

Functional properties of helical **HELs** were efficiently utilized as a supramolecular chiral unit in molecular sensing devices to sense and discriminate optically active compounds. Indeed, homochiral **HELs** are able to stereoselectively interact with another enantiopure molecules to form a supramolecular complex. This results in the chirogenic process of asymmetry transfer/modulation from a chiral guest to **HEL** through the corresponding non-covalent interactions, which can be conventionally monitored by various spectroscopic techniques. In this respect, many scientific reports and reviews have been published from different research groups in the recent years,[6f,46,47] and here we discuss the most interesting representative results.

Initially, non-racemic **[6]HEL** was used to test the chirality of non-racemic amine by enantioselective quenching of fluorescence, and it was found that the quenching behavior was highly dependent on the polarity of the solvent but not upon the asymmetry of the amine itself (while if present, only small and within experimental error).[48a] Further, the hydroxyl substituents have been added to the **HEL** structure to obtain 2,15-dihydroxy-[6]helicene (**HELIXOL**), which was used as an enantioselective fluorescence sensing probe for chiral amines and amino-alcohols (Figure 10).[48b] In this study, (***P***)-(**+**)-/(***M***)-(**-**)-**HELIXOL** displayed enantioselective discrimination observed by fluorescence quenching upon complexation with the corresponding chiral analytes. Among the chiral analytes screened by levorotatory (***M***)-**HELIXOL**, alaninol shows the best result with the enantioselective factor $\alpha = K_R/K_S = 2.1$. The presence of two phenolic hydroxyl groups in the periphery of (***M***)-**HELIXOL** ensures a more tightly bound 1:1 host:guest hydrogen-bonded enantioselective complex with amino alcohols than that with amines, thus exhibiting a stronger fluorescence quenching and high enantioselectivity. The existing hydrogen bonding interaction provides an optimal steric

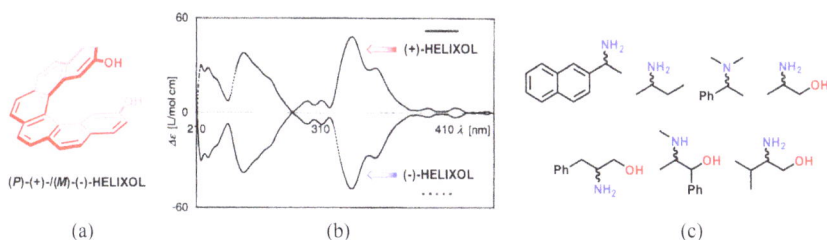

(a)   (b)   (c)

Figure 10. (a) Structure of "**HELIXOL**", (b) CD spectra of (***P***)-(**+**)-**HELIXOL** and (***M***)-(**-**)-**HELIXOL**, Reprinted with permission from Ref. [48b]. Copyright 2001, Elsevier, and (c) HELIXOL enantioselective recognition of chiral amines and amino-alcohols monitored by fluorescence quenching.

influence for the chiral recognition of alaninol in the inner core of (*M*)-**HELIXOL** to secure the efficient chirogenic process.

Further, to increase the chiral recognition ability of **HELs** by tethering a crown ether part (**Me₄-[5]HEL-CE** and **[6]HEL-CE**), which connects the upper and lowers terminal rings, the corresponding crown ether derivatives have been obtained. This modification provides a substantial contribution toward the chiral discrimination of racemic α-amino organic ammonium salts (α-**AOA**) in an enantioselective liquid–liquid (organic-aqueous) extraction process (Scheme 10).[49] In particular, chiral HELs were able to selectively extract the respective α-**AOA** ions (α-**AOA-COOMe, α-AOA-Ph, α-AOA-Me**) from an aqueous to organic layer (CHCl₃) and form the diastereoisomeric complexes. These diastereomeric complexes displayed two distinct -COOMe (in α-**AOA-COOMe**) ¹H-NMR signals and its enantioselectivity, extraction parameters, and optical purity were calculated by using UV–Vis absorption data of the

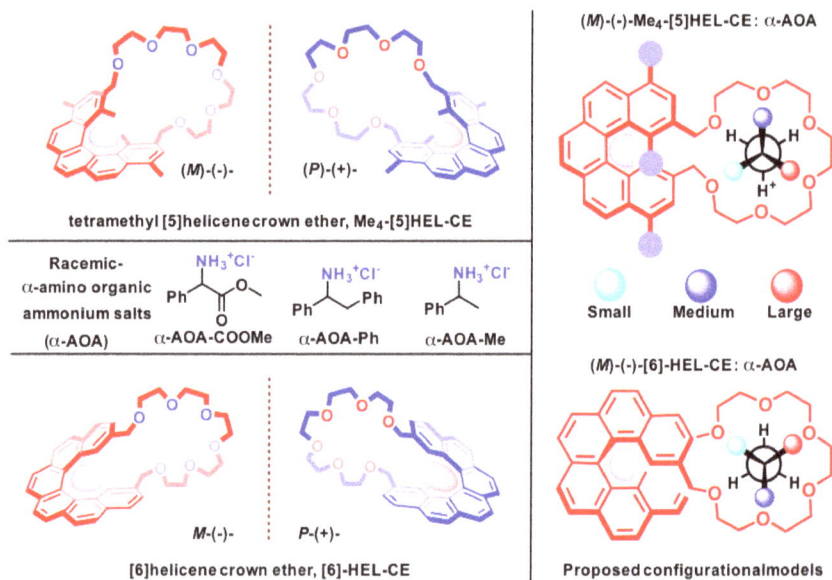

Scheme 10.   Enantioselective extraction of chiral α-amino organic ammonium salts, α-**AOA** from aqueous to organic layer using homochiral [5]- and [6]helicene-tethered crown ether derivatives, **Me₄-[5]HEL-CE** and **[6]HEL-CE**. Proposed configurational models for more efficient binding enantiomers of α-**AOA** with (*M*)-(-)-**Me₄-[5]HEL-CE** and **[6]HEL-CE**.

reverse extraction of complexed guest back to an aqueous acidic phase. In this enantioseparation process, **Me₄-[5]HEL-CE** exhibited better enantioselective extracting ability (extraction 6%, optical purity 75%, 6 h) than the **[6]HEL-CE** (extraction 2%, optical purity 26%, 6 h). Furthermore, the methyl groups of the peripheral ring at the upper and lower sides of **Me₄-[5]HEL-CE** provide considerable steric influence for the efficient chirogenic process with chiral selectivity.

(*M,M*)-[5]HELOL is a **HEL**-based biaryl diol molecule that actively reacted with PCl_3, *in situ*, to generate a chloro phosphite derivative was used as a chiral ^{31}P NMR derivatizing agent to discriminate chiral alcohols and amines acting as nucleophilic guests (Scheme 11).[50a] This supramolecular host was used to sense and discriminate the remote chirality of a molecule, where the chiral center locates far away (for example, 8-phenylnonanol and vitamin E) from the reacting phosphite group. Further, the chirogenic ability of **HELOL** was enhanced by additional structural modification leading to the development of remote chirality sensors for variety of alcohols, phenols, amines, and carboxylic acids (Scheme 11).[50b]

Recently, Yang and co-workers developed an efficient hybrid supramolecular chirogenic system, chiral 3-aza[6]helicene tethered β-CDX, **(*P*)-(+)-/(*M*)-(-)-[6]HEL-CDX**, for chiral recognition of underivatized proteinogenic amino acids in an aqueous medium (Scheme 12(a)).[32]

This modified hybrid system containing a chiral cavity shows significantly high chiral discrimination for leucine and tryptophane at pH 7.3 (Scheme 12(b)). The efficient binding affinity and high enantioselectivity of amino acids were achieved by the synergetic chirogenic effect of self-included **3-aza[6]HEL** to hydrophobic CDX cavity. In this hybrid system, the combined effect of the intrinsic chiralities (helical and non-specific hydrophobic chirality, respectively) of the two components were fine-tuned and provided a confined chiral selective pocket for the co-included amino acid.

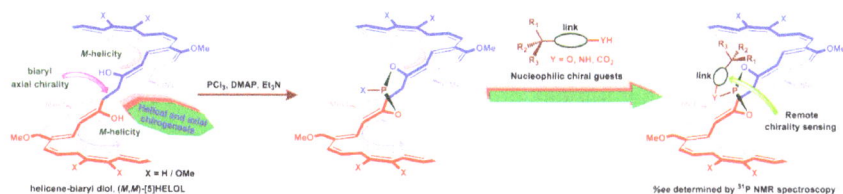

Scheme 11.   Sensing of enantiopure alcohols, phenols, amines, and carboxylic acids using helicene-biaryl diol, **(*M,M*)-[5]HELOL** by ^{31}P NMR measurements.

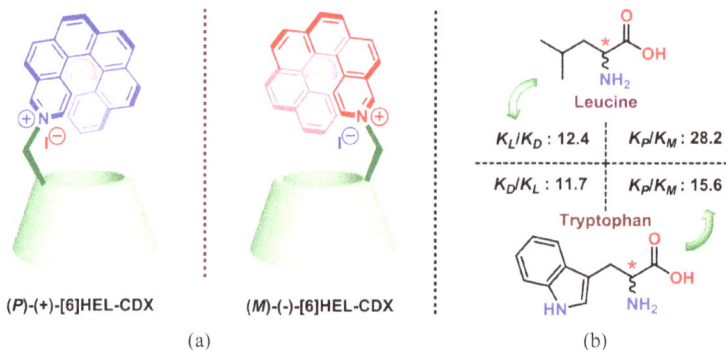

Scheme 12. (a) Structure of **(P)**-(+)- and **(M)**-(-)-3-azonia[6]helicenyl β-CDXs, **(P)**-(+)-/**(M)**-(-)-**[6]HEL-CDX**, and (b) its selectivities toward underivatized proteinogenic amino acids in aqueous solution at pH 7.3.

The helically grooved sensors and the chiral discrimination methods significantly contributed to the development of efficient supramolecular chirogenic systems, especially upon using suitably modified helicene-based hybrid hosts.

3.1.4. *Supramolecular self-assembling systems*

On the other hand, supramolecular chirogenic sensing and recognition is not only a process of the interaction between a host molecule and antipodal guest molecules but also deals with the differential interaction behavior of complex supramolecular systems (self-assemblies, helical polymers, nanoscale structures) with chiral molecules. The chiral recognition of self-assemblies is one of the important applications. These chiral interactions may lead to new functionalities and useful features, which can be effectively applied to monitor the chirality excess of chiral components. Thus, the most important and advanced characteristics of chiral recognition-induced observable changes in supramolecular self-assemblies are rheological properties (solution–solid, sol-gel transformations), visible/naked eye color change, formation of macroscopic chiral nanoscale structures, easily detectable spectroscopic response (UV–Vis and CD spectra, etc.), and so on.[2b,10c] The efficiency of supramolecular chirogenesis in self-assembled systems is strongly affected by the involved reversible and divergent non-covalent interactions between the components present, the medium used (solvents, temperature, ionic strength, pH, etc.), and the external stimulus (photoirradiation, redox effect, chemical additives, sonication, circularly polarized light, etc.).[2b,10c] Taking advantage of chiral supramolecular self-assemblies, these systems were further successfully used in diverse applications like asymmetric catalysis, chiroptical switches, optics, and electronics-based functional devices, white light-emitting CPL polymers, preparation of functional soft-materials etc.[2a,b,10c,21a]

First, Pu *et al.* reported a selective, naked eye chiral differential recognition by a supramolecular self-assembled organogel system, **(R)-GA1**, which changed its physical sol-gel state upon complexation with different antipodal amino alcohols (including prolinol, valinol, phenylalanine, leucinol, and 1-amino-2-propanol).[51] The chiral gelator contained (R)-1,1'-bi-2-naphthol conjugate-tethered terpyridine Cu(II) complex, **(R)-GA1** (15 mg), which formed a stable supramolecular gel in chloroform upon sonication and remained stable upon addition of (R)-phenylglycinol, whereas the gel collapsed upon addition of (S)-phenylglycinol under the same condition (Figure 11(a)). Similarly, a relatively weak fluorescence emission of **(R)-GA1** at 396 nm was enormously increased with (S)-phenylglycinol, whereas with (R)-phenylglycinol only much weaker fluorescence enhancement (Figure 11(b)) was observed. This visible, chirogenic response of sol-gel transformation was attributed to the enantioselective displacement of Cu(II) ion from the **(R)-GA1** gel by amino alcohol. This unprecedented enantioselective gel collapsing of the chiral supramolecular gel is a potentially useful system for visual discrimination of chiral components.

Similarly, another visually detectable enantioselective sensing system has been reported for the differentiation of (R)-/(S)-(1,1'-binaphthalene-2,2'-diyl)*bis*(diphenylphosphine), (R)-/(S)-BINAP derivatives using a metalloregulator, **GA2** and **GA3**, system containing a pincer-type platinum complex tethered with cholesterol fragment (Figure 12(a)).[52] These metallogelators, **GA2** in chloroform (Figure 12(b)) and **GA3** in toluene, formed stable metalogels with (R)- and (S)-BINAP through the corresponding $\pi-\pi$ stacking and metal–metal interactions. In most cases, these gels collapsed by subsequent heating to reflux and cooling to room temperature. However, surprisingly, 0.1 equivalent of (S)-BINAP enantiomer containing metalogel **GA2** appeared as a stable gel, being robust with heating and cooling sequences, whereas 0.1 equivalent of (R)-BINAP enantiomer containing metalogel was collapsed under the same conditions

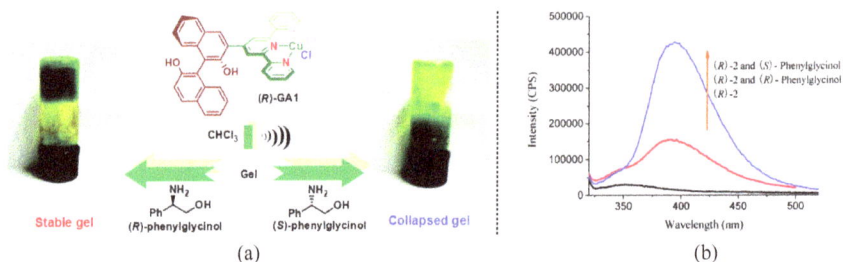

(a)   (b)

Figure 11.   (a) Schematic representation of naked eye, enantioselective sensing response of chiral gel, **(R)-GA1** toward (R)- and (S)-phenylglycine in chloroform, (b) fluorescence spectra of **(R)-GA1** (5.0×10^{-7} M) in CH_2Cl_2/n-hexane (2:3) in the presence of (R)- and (S)-phenylglycinol (5.0×10^{-4} M) (λ_{exc} = 289 nm, λ_{emi} = 396 nm slits: 2 nm/5 nm). Reprinted with permission from Ref. [51]. Copyright 2010, American Chemical Society.

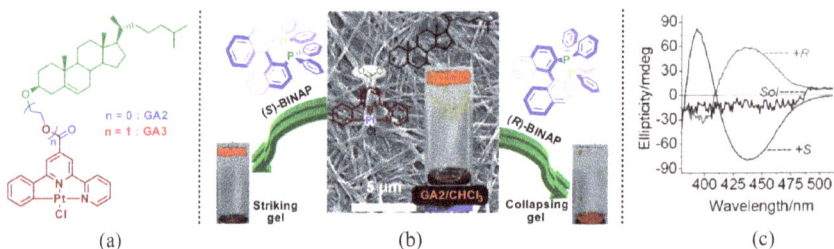

(a)   (b)   (c)

Figure 12.   (a) Structure of metallogelators, **GA2** and **GA3**, (b) visually detectable chiral recognition of (R)- and (S)-BINAP through enantioselective metallogel **GA2** striking-collapsing respectively, and (c) Sol: **GA2**/CHCl$_3$ (0.125 wt%); +R: solution prepared with **GA2**/CHCl$_3$ (0.125 wt%) and 1 equiv. (R)-BINAP; and +S: solution prepared with **GA2**/CHCl$_3$ (0.125 wt%) and 1 equiv. (S)-BINAP. Reprinted with permission from Ref. [52]. Copyright 2011, Wiley-VCH Verlag GmbH & Co. KGa.

(Figure 12(b)). The observed stereoselectivity of metallogel **GA2** was due to the matching–unmatching structures of the corresponding enantiomers with the chiral environment created by cholesterol fragments and the coordination of bulky binaphthalene skeleton attached PPh$_2$ groups to Pt center that caused blocking the π–π stacking and metal–metal interactions.

Another chiral supramolecular gelator, **GA4**, contained achiral zinc tetraphenylporphyrin functionalized with an enantiopure gelator unit on the basis of alkylated *L*-glutamate. This system formed highly ordered

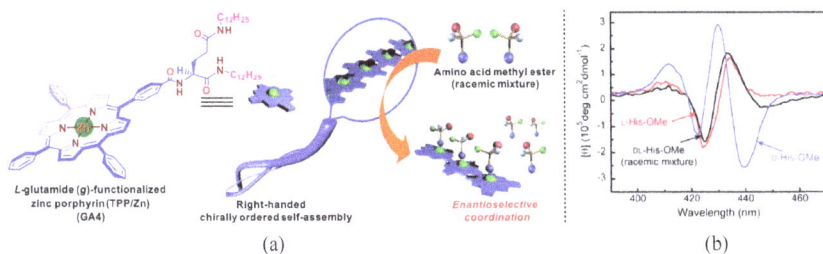

Figure 13. (a) Schematic illustration of the enantioselective recognition of amino acid methyl esters through chiral ordered porphyrin-based self-assembly of **GA4**. (b) CD spectra of **GA4** (50 mM) with *D*-His-OMe (200 mM), *D,L*-His-OMe (racemic mixture, 400 mM) and *L*-His-OMe (200 mM) in cyclohexane at 20°C. Reprinted with permission from Ref. [53b]. Copyright 2012, Royal Society of Chemistry.

chiral self-assemblies in organic solvents and showed sensitive chiroptical response to axial coordination with basic ligands[53a] and exceptionally high enantioselectivity with amino acid methyl esters (Figure 13).[53b] Among the tested compounds, *L*- and *D*-histidine derivatives exhibited different CD patterns and fluorescence quenching-based Stern–Volmer constants with the enantioselectivity (K_{SV} (*L*)/K_{SV} (*D*)) of 3.74. The exceptionally high enantioselectivity and sensitivity of **GA4** with these amino acid derivatives were attributed to a change in the assembling structures and chirally ordered stacking of the binding sites but not to molecular chirality.

3.2. *Supramolecular photochirogenic reactions*

A particular branch of chirogenesis dealing with light-induced chirality transfer is called photochirogenesis that is able to create new stereogenic centers and a divergent range of new chiral photochemical products with high enantiomeric excess. Even though vigorous progress and development in the photochirogenesis research in past decades,[6c,7c,54] controlling the stereochemistry of light-induced reactions has been theoretically rationalized but not completely achieved yet by experimental methods and conditions, this is a result of inefficient chirality transfer occurring in the electronically excited states rather than in the ground state. Besides, it creates a very small energy gap difference in the diastereomeric transition states that govern the enantiomeric photoproducts.[6b] Since many photochirogenic strategies have been proposed and reported,[7c,54a] supramolecular systems (see Section 2) mediating the photochirogenic reactions are highly fascinating and paradigmatic challenging task for many researchers to attain high selectivity and enantiomeric excess of the formed photoproducts by manipulation of the non-covalent interactions (e.g., hydrogen bonding, Van der Waals, coulombic, hydrophobic, and $\pi-\pi$ interactions) involved between the host and guest. The efficiency of photochirogenesis was also controllable by tuning some internal and external crucial factors like chiral inductors, temperature, solvent, and pressure. Up to now, the results of supramolecular photochirogenesis strategies have emerged as a promising tactical intelligence, as it benefits from relatively long and strong supramolecular interactions and its synergetic confinement effects exerting their influence on both the ground and electronically excited states.[2a,6b,c,10a,b,55] These combined effects of supramolecular systems distinctively increase the activation energy gap of the electronically excited pair of diastereomeric intermediates or transition states and enhances the photoproducts enantioselectivity.

So far, a variety of supramolecular hosts such as CDX derivatives, chiral metal cages, chiral hydrogen bonding templates, chiral auxiliaries, chirally modified zeolites, bio-macromolecules, and chiral aggregates, etc., possessing a chiral cavity/pocket have been effectively employed as supramolecular chiral mediators for many stereoselective photoreactions to attain high enantioselectivity. While there are several comprehensive reviews in this field,[2a,6b,c,10a,b,55] in this part we focus on a few essentially crucial and recent scientific reports that deal with the advances of photochirogenesis in the different supramolecular system.

3.2.1. *In CDXs*

As discussed above (see Section 3.1.1), naturally occurring CDXs are inherently chiral, serving as prominent supramolecular hosts for various chirogenic events. These widely employed CDXs possess such important advantages as good aqueous solubility, optically transparency over the whole UV–Vis region, and feasibility of chemical modification with divergent functional groups. Furthermore, it can accommodate a wide range of organic guest molecules in the hydrophobic cavity with the tunable multiple supramolecular interactions. Owing to these beneficial features, CDXs can effectively serve as chiral differentiating agents (see Section 3.1.1),[26a] chirogenic supramolecular host for photochemical reactions(6b-d), and chiral catalyst for many organic transformations.[56]

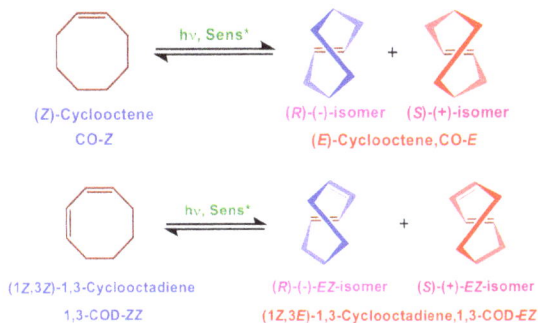

Scheme 13. The enantiodifferentiating photoisomerization of cyclooctene, **CO-Z**, and 1,3-cyclooctadiene, **1,3-COD-ZZ**.

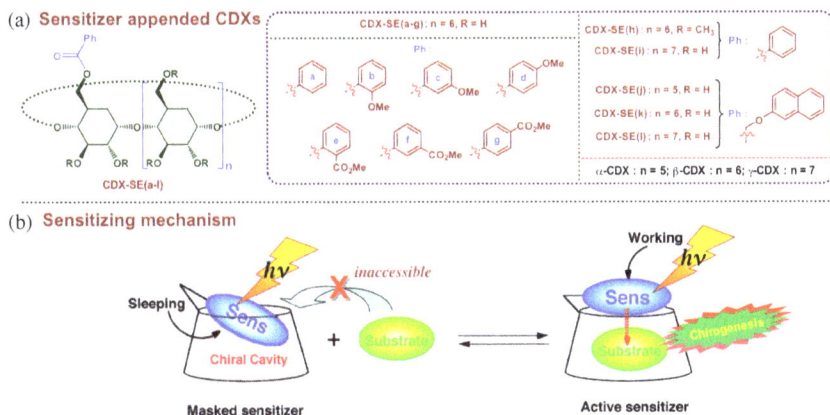

Figure 14. (a) Structure of sensitizer appended α-, β-, and γ-CDXs, **CDX-SE(a-l)** for supramolecular photosensitized chirogenic reactions, and (b) chirogenic sensitization mechanism of **CDX-SEs**. Reprinted with permission from Ref. [6b]. Copyright 2014, Royal Society of Chemistry.

One of the earliest and widely studied authentic photochemical reactions is enantiodifferentiating photoisomerization of (Z)-cyclooctene, **CO-Z** mediated by native and modified CDXs (Scheme 13). Direct irradiation (185 nm) of solid-state 1:1 complex of β-CDX:**CO-Z** affords nearly racemic (E)-cyclooctene, **CO-E** with the *ee* value of only 0.5–1.5%. This poor photochirogenic ability of native β-CDX was enhanced up to 11% *ee* (in aqueous solution) by a catalytic amount (0.1 equiv.) of the sensitizer, benzoyl-modified CDX (Figure 14).

To date, a wide range of sensitizer-tethered CDXs (α-, β- and γ-CDXs) having different sizes of the chirogenic cavities have been prepared and its photochirogenic efficiencies have been investigated in the photoisomerization of **CO-Z** and 1,3-cyclooctadiene, **1,3-COD-ZZ**.[57,58] These sensitizer-tethered CDXs, **CDX-SE(a-l)** provided efficient photochirogenesis due to the self-inclusion of a hydrophobic sensitizer moiety (which was also fine-tuned and modified the chiral confinement of CDX cavity), which was expelled to the portal area upon guest binding, hence making a close contact with each other and subsequently facilitating the light-triggered energy transfer process to initiate the chirogenic photoreactions.

Mainly, the cavity size of CDXs, nature and position of substituent(s) in the benzoate sensitizers, **CDX-SE(a-l)**, and various external factors (like solvent and temperature) are critical for supramolecular sensitization.[57,58] As mentioned before, parent β-CDX failed to produce a noticeable *ee* in the enantiodifferentiating photoisomerization of **CO-Z**.[57b,c,59a] However, the reaction efficiency was considerably enhanced up to 11% for 6-*O*-benzoyl-β-CDX, **CDX-SE(a)**,[59b] to 24% for 6-*O*-(methyl phthaloyl)-β-CDX, **CDX-SE(b)**,[57b] and even to 46% for 6-*O*-(*m*-methoxybenzoyl)-β-CDX, **CDX-SE(c)**.[58a,b] Further, a similar though less pronounced effect was observed for **1,3-COD-ZZ** with β-CDX and corresponding analogs, **CDX-SE(j-l)**.[57c]

Further, the efficiency of supramolecular photochirogenic systems was improved by fine-tuning the chiral environment and confinement of the host system in different architectural nanosponges,[60] gels,[61] and liquid crystals.[62] Interestingly, the *ee* value of **CO-E** was more dramatically controlled by changing the phase from sol-gel, pyromellitate cross-linked nigerosylnigerose nanosponges (CNN-NSs). Low *ee* values has obtained in solution and suspension phases and has enhanced up to 22–24% upon gelation.[60c] This is the highest *ee* value obtained in the case of CDX-based nanosponges (CD-NSs) (6–12%).[60b] However, upon gelation of pyromellitate cross-linked linear maltodextrin-based nanosponges (LM-NSs), the *ee* value experienced a sudden drop.[60d] These sol-gel results clearly indicate that the photochirogenesis of host cavity in CD/CNN-NS plays a crucial role in the enantiodifferentiation of **CO-Z** in comparison to the chiral void space created upon gelation of LM-NSs.

Recently, Yang and co-workers reported the [4]rotaxanes, **Rot-CPh** and **Rot-CNp** mediated photoisomerization of **1,3-COD-ZZ** to **1,3-COD-EZ** affording enantiodifferentiation up to 15.3% (in **Rot-CNp**), which is the highest value for this reaction achieved so far by

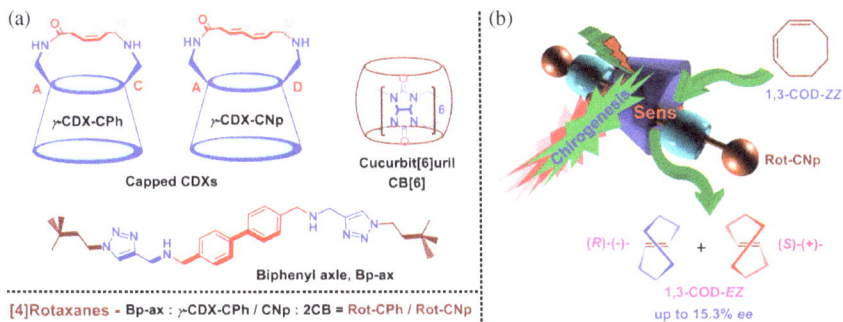

Scheme 14. (a) Structure of capped CDXs, **CDX-CPh** and **CDX-CNp**, CB and the axle in the [4]rotaxane, **Rot-CPh** and **Rot-CNp**, and (b) the [4]rotaxane, **Rot-CNp**-mediated photoisomerization of **1,3-COD-ZZ** to **1,3-COD-EZ**.

supramolecular photochirogenesis (Scheme 14).[63] These rotaxanes contain a biphenyl photosensitizer axle and azide stopper. The axle was encapsulated with native and capped γ-CDXs (**CDX-CPh** and **CDX-CNp**) and templated by rotaxanation of the CB[6] unit through the azide stopper, which was obtained *via* azide-alkyne 1,3-dipolar cycloaddition. In this strategy, rotaxanation of capped γ-CDXs containing rigid aromatic rings and interacting with biphenyl axle makes the CDX hydrophobic cavity a highly confined chiral binding and sensitizing site for **1,3-COD-ZZ** at both the ground and excited states with the chirogenesis occurring efficiently.

Another extensively investigated supramolecular photochirogenic reaction is enantiodifferentiating [4+4] photocyclodimerization of 2-anthracenecarboxylate (AC)[64a,b] to give classical cyclodimers, **CC1-CC4** and non-classical/slipped cyclodimers, **NC5** and **NC6** (Scheme 15). Thus, native γ-CDX forms a 1:2 host–guest complex (stepwise association constants of $K_{1:1}$ = 161 M^{-1} and $K_{1:2}$ = 38,500 M^{-1}) affording classical photocyclodimers, *syn*-HT **CC2** in 41% *ee*, and *anti*-HH **CC3** in <5% *ee* at 0°C,[64c] whereas native β-CDX forms a 2:2 host–guest complex (stepwise association constants of $K_{1:1}$ = 3800 M^{-1} and $K_{2:2}$ = 150 M^{-1}) affording slipped photocyclodimers of *anti*-HT **NC5** in 47% *ee*, *syn*-HT **NC6** in 24% *ee* along with classical achiral *anti*-HT **CC1** in the relative yield of 65% in aqueous buffer at 25°C (Scheme 15).[65]

Up to now, a wide variety of the modified CDXs derivatives with different functional groups have been prepared by selective and

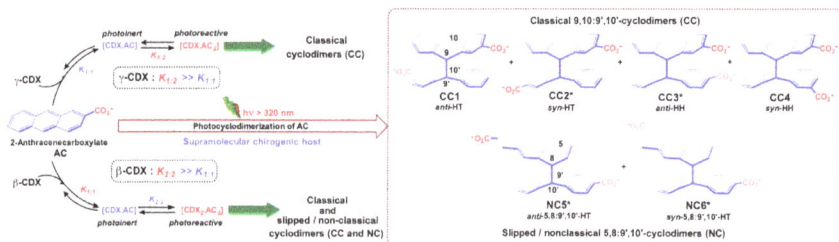

Scheme 15. Supramolecular chiral host (native γ- and β-CDXs)-mediated enantio-differentiating photocyclodimerization of 2-antharacenecarboxylate (AC) to give classical 9,10:9′,10′-cyclodimers **CC1-CC4** and slipped/non-classical 5,8:9′,10′-cyclodimers **NC5** and **NC6**.

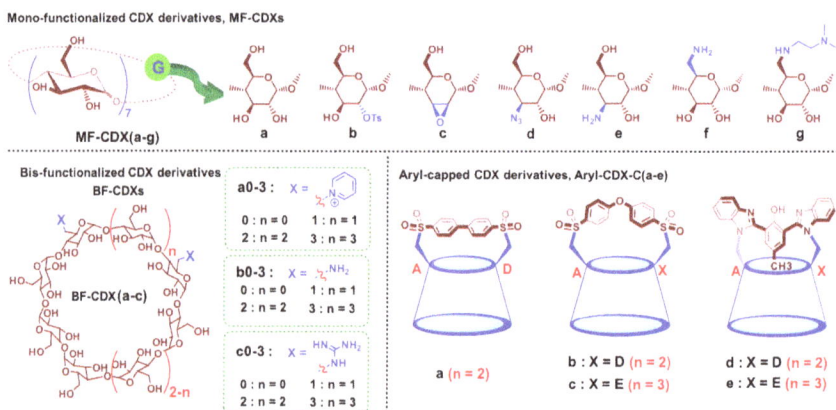

Figure 15. Examples of selectively functionalized CDXs derivatives, **MF-CDXs**, **BF-CDXs** and **Aryl-CDX-Cs**, as supramolecular photochirogenic hosts for enantio-differentiating photocyclodimerization of AC.

per-functionalization of primary and secondary sides, **MF-CDXs**, **BF-CDXs,** and **Aryl-CDX-Cs** (Figure 15) and used as supramolecular photochirogenic hosts to enhance the chemical and optical yields of the photoproducts of AC.[66,67] The selectively functionalized CDX derivatives are able to greatly accelerate the photocyclodimerization of AC and achieve the high chiral product distribution of *anti/syn-*, HT/HH- ratios, and enantioselectivities.[66,67] Besides, the product distributions and enantioselectivities were also critically affected by various external factors such as solvent, temperature, pressure, irradiation wavelength, and additives (like metal salts, etc.).[6b–d,66,67]

Recent investigations have been mainly focused on the efficiency of photochirogenesis, while the effect of chirality inversion was achieved by addition of co-solvents (as ammonia) and some additives and changing the irradiation wavelength. For example, selective introduction of two amino-functional groups in the primary rim of γ-CDX (**BF-CDX(b)**) ensured the attractive electrostatic interactions with the included AC and comparatively enhanced the yield and *ee* of photoproduct, HH dimer **CC3**, up to 22% and −27%, respectively, in aqueous methanol at −45°C,[67b] whereas native γ-CDX afforded only ~8% yield and <5% *ee* in a pH 9 aqueous buffer solution at 25°C.[64c]

Despite the wide range of employed modified CD hosts, the mediated enantiodifferentiating photocyclodimerization of AC gives the respective photochirogenic outcomes, which were also altered by changing the solvents at various temperatures. Interestingly, the enantiodifferentiating photocyclodimerization of AC mediated by $6^A,6^E$-diguanidino-γ-CDX, **BF-CDX(c3)** as a supramolecular photochirogenic host also showed the chirality inversion and enhancement of photoproduct, *anti*-HH dimer **CC3**, with the yield and *ee* values being 72% and −86%, respectively at −85°C in a 80% aqueous NH$_3$ solution and 46% and 64%, respectively at −70°C in 60% aqueous methanol.[68] In contrast, diamino-γ-CDX host, **BF-CDX(b3)** did not show such unusual photochirogenic behaviors as chirality inversion and enhancement of photoproduct. The diguanidino-γ-CDXs, **BF-CDX(c0-2)**-mediated enantiodifferentiating photocyclodimerization of AC also greatly enhanced the yields of HH photo dimers, **CC3,** and **CC4**, while the enantioselectivities of chiral photodimers **CC2** and **CC3** were also inverted upon changing the external factors like temperature and solvent composition.[69] The observed enantiodifferentiating results of photocyclodimerization of AC mediated by **BF-CDX(c0-3)** showed that the *ee* value of the chiral photodimers is increased upon lowering the temperature and the *ee* sign is dependent upon the ammonia content to reveal the opposite temperature dependencies at low and high ammonia contents (10–80%). This is due to the fact that the altered solvent structure and guanidinium–carboxylate interaction mode are responsible for the effective photochirogenic outcome and inversion of HH photo dimer **CC3**.

As the photocyclodimerization of AC is an excited-state reaction, the *re*/*si*-enantiotopic face selectivity of the photoproducts, was manipulated

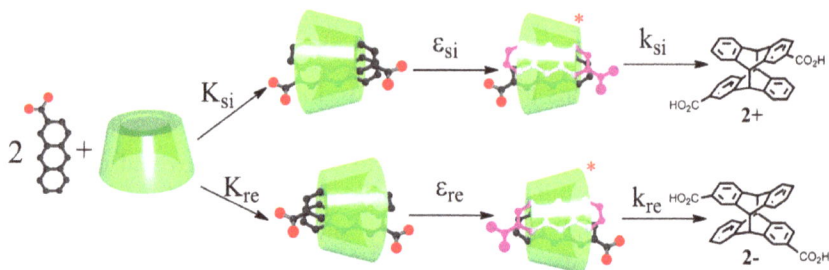

Scheme 16. Manipulating γ-CDX-mediated photocyclodimerization of AC by wavelength, temperature, solvent, and host. Reprinted with permission from Ref. [70a]. Copyright 2014, ScienceDirect, Elsevier.

or decided in the ground state through complexation with CD (as 1:1 or 1:2) (Scheme 16).[70a] Further, the stereochemical outcomes of photoproducts (*syn/anti*, HT/HH ratios, and *ee* values) were altered by the host structure and its chirogenic effects, and also critically dependent on the irradiation wavelength.[70b] The employed host structure, reaction temperature, and used solvent affected the ground-state stacking geometry of the complexed AC pair in the CD cavity, while the photoreaction carried out in the optimized irradiation wavelength was able to achieve high stereochemical outcome of the chiral photoproducts (for example, *syn*-HT, **CC2** in 54% *ee* and *anti*-HH dimer, **CC3** in −37% *ee*).

Besides the various controlling factors described above, it is important to note that metal ions may play a key role in the photochirogenic processes. Thus, Inoue *et al.*, reported the first catalytic enantiodifferentiating photocyclodimerization of AC mediated a metallosupramolecular photochirogenic host/catalyst, diamine side chain-tethered γ-CDX-Cu^{2+} complex, γ-**CDXDA-Cu^{2+}** (Scheme 17(a)).[71] According to this strategy, the photochirogenesis efficiency was considerably enhanced up to 64–70/43% *ee* and 51–52/43% yield of *anti*-HH dimer **CC3** by using a 0.1/0.01 equiv. of γ-**CDXDA-Cu^{2+}** as a catalyst. Recently, Yang and co-workers reported the first visible-light-driven TTA-based catalytic photochirogenic enantiodifferentiating photodimerization of AC affording 60.8% conversion of *syn*-HT dimer, **CC2** in 31.4% *ee* in the presence of 0.5% equiv. of the photocatalyst (Scheme 17(b)).[72a] In this catalytic system, the host, γ-CDX tethered with the Pt(II) Schiff base sensitizer, γ-**CDX-Pt(II)**, hence absorbing the visible light and transferring the excited state energy *via* the Dexter mechanism[72b] to two bound AC molecules inside the CDX

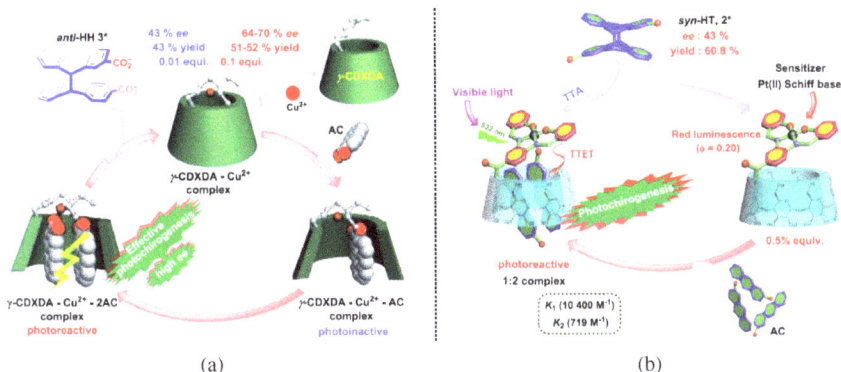

(a)

(b)

Scheme 17. Catalytic enantiodifferentiating photocyclodimerization of AC mediated a (a) metallosupramolecular photochirogenic host, γ-**CDXDA-Cu^{2+}**, Reprinted with permission from Ref. [71]. Copyright **2009**, Wiley-VCH Verlag GmbH & Co. KGaA, Weinheim, and (b) Pt(II) Schiff base sensitizer-tethered γ-CDX, γ-**CDX-Pt(II)** mediated visible-light-driven triplet–triplet annihilation (TTA)-based catalytic photochirogenesis. Reprinted with permission from Ref. [72a]. Copyright 2018, American Chemical Society.

cavity followed by the photochirogenic reaction and expulsion of the photoproducts from the cavity to repeat the photocatalytic cycle.

3.2.2. *In self-assembled metal cages*

Irrespective of many supramolecular chirogenic hosts and templates employed for photochirogenesis, a variety of self-assembled metal cages having chirogenic cavities have been utilized and emerged recently. In particular, self-assembled metal-organic coordination frameworks create voids/cages-like cavities that can be used as molecular containers/flasks with remarkable functions such as molecular recognition, adsorption of toxic gases, interaction with organic molecules and metal ions, stabilization of reactive species, catalysts/additives/hosts for various stereoselective thermal and photochemical reactions, and others.[67a,73,74]

For example, Fujita *et al.*, utilized the water-soluble self-assembled M$_6$L$_4$ coordination cages (Figure 16) for stereoselective photodimerization of olefins, Diels–Alder reactions, and asymmetric [2+2] photocycloaddition reactions.[75–79] These self-assembled Pd(II) coordination cages, Pd-CG (Scheme 18(a)), created a confined microenvironment for the included various reactant molecules, which undergo photochirogenic

(a)

(b)

$\Lambda\Lambda\Lambda\Lambda$-(Me₄N)₄[Fe₄L₆] $\Delta\Delta\Delta\Delta$-(Me₄N)₄[Fe₄L₆]

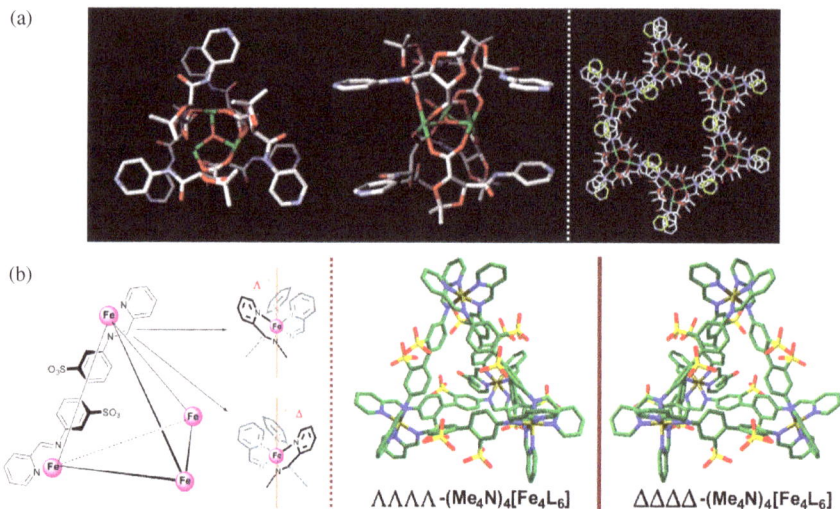

Figure 16.   Structure of (a) self-assembled trinuclear metal cage POST-1. Reprinted with permission from Ref. [80]. Copyright 2000, Nature Publishing Groups, and (b) tetrahedral metal cage $[Fe_4L_6]^{4-}$. Reprinted with permission from Ref. [81]. Copyright 2016, Science Direct, Elsevier.

reaction in the excited state and subsequently yield the photoproducts with good to excellent regio- and stereoselectivity. By introducing chiral ligands to the periphery of Pd-CG cage (**Pd-CG-Me/-HMe/-HEt/-HH**), the inside hydrophobic electron-deficient cavity becomes chiral and the photocycloaddition of fluoranthenes to *N*-cyclohexylmaleimide derivative in the cavity of **Pd-CG-HEt** gives photoproduct (R = Me) with 50% *ee*.[76] The asymmetric chirogenic induction of photoproducts (R = H) is sensitive to the nature (steric bulk) of substituents on *N* in externally introduced chiral auxiliary ligands (Scheme 18(b)). The absolute configuration of photoproduct R = H was confirmed by X-ray crystal structure (Scheme 18(c)).

In a related study, homochiral mesoporous metal-organic materials, POST-1, have been used as a supramolecular chirogenic host for the enantio-differentiating photoisomerization of **CO-Z** to give **CO-E** in 5.4% *ee*. The observed low enantioselectivity was due to the large cavity size (diameter 13 Å) of chiral channels in POST-1 with the created microenvironment being inefficient to deliver the chiral information to **CO-E** (Figure 16(a)).[67a,80]

Recently, Yang and co-workers reported the iron tetrahedral metal cages, $[Fe_4L_6]^{4-}$ as a supramolecular chirogenic host for the enantiodifferentiating photocyclodimerization of AC in an aqueous medium (Figure 16(b)).[81] The

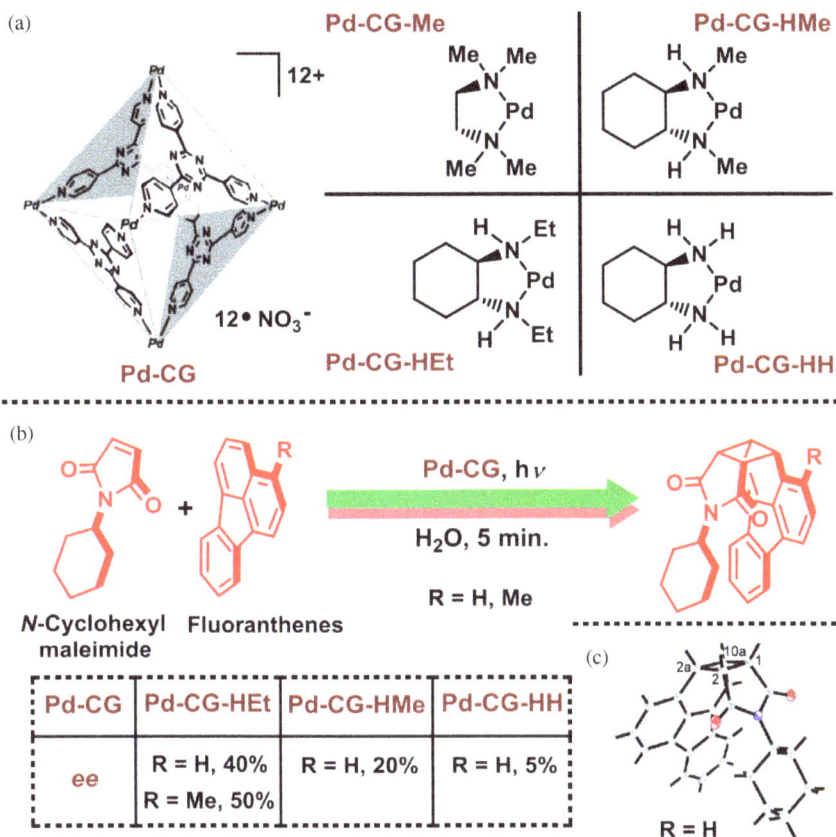

Scheme 18. (a) Self-assembled M_6L_4 Pd(II) coordination cages, Pd-CG, (b) photoreaction of fluoranthenes in the cavity of Pd-CG, and (c) X-ray structure of the photoproduct (R = H) (30% probability level). Reprinted with permission from Ref. [76]. Copyright 2008, American Chemical Society.

racemic iron tetrahedral metal cages enhanced the amount of *anti*-HT photo-dimer, **CC1**, whereas the enantiopure metal cage produced moderate enanti-oselectivity of chiral *syn*-HT (**CC2**) and *anti*-HH (**CC3**). Further, external factors, such as solvent and temperature have been found to play a remark-able role (accommodation of AC into the flexible reaction sites at the periph-ery of metal cage rather than its rigid cavity) on the photoproduct distribution and enantioselectivity. The metal cages formed in a stepwise manner 1:1 and 1:2 complexes with AC ($K_{1:1}$ = 2350 M^{-1} and $K_{1:2}$ = 580 M^{-1}), *via* a partial insertion of the first AC molecule into the hydrophobic cage, while the sec-ond AC interacts with the periphery side of the cage through π–π stacking with the biphenyl moiety. From the above results and obtained moderate enantioselectivity, it is likely that the reaction occurred at the periphery of metal cage rather than inside the rigid cavity.

3.2.3. *Other miscellaneous chiral supramolecular systems*

In addition to the above described supramolecular chirogenic systems, several other hosts are important to discuss owing to their interesting properties that have been effectively used for enantiodifferentiating isomerization of **CO-Z** and cyclodimerization of AC.

Thus, Inoue and co-workers reported the planar-chiral paracyclophanes, (*R*)-**CPs** as efficient sensitizers being a planar-to-planar chirogenic host for the enantiodifferentiating photoisomerization of **CO-Z** and (*Z,Z*)-1,5-cyclooctadiene **1,5-COD-ZZ** (Scheme 19).[82] The decamethylene bridge in planar-chiral paracyclophanes bound with the substrates **CO-Z/1,5-COD-ZZ** and effectively shielded one of the enantiotopic faces upon sensitization, while its planar chirality was delivered in the excited state and yielded the enantiotopic face-selective photoproducts, **CO-E/ COD-EZ** with high *ee*. The planar-to-planar chirality varied with respect to the substituent group R in **CP**: **CP(a)** and **CP(b)** and afforded **CO-E** in 42% and 43% *ee* at −110°C and **1,5-COD-EZ** in 87% and 85% *ee* at −140°C, respectively.

PAs are another interesting macrocyclic supramolecular host having the corresponding cavities to accommodate various organic guests of a suitable size mainly through the electrostatic dipole interactions and C–H–π/π–π binding in organic solvents. However, these **PAs** can be converted into the water-soluble structures by the introduction of suitable functional groups (cationic or anionic) on both sides of the peripheral rims.[83,84] Thus, water-soluble cationic pillar[6]arenes, **WPA[6]** have been

(Z,Z)-1,5-Cyclooctadiene   (*R*)-(-)-isomer   (*S*)-(+)-isomer   Sens* :   (CH₂)₁₀
1,5-COD-ZZ   (*EZ*)-1,5-Cyclooctadiene, 1,5-COD-*EZ*   ROOC   (*R*)-CP
R = Me / *i*-Pr : CP(a) / CP(b)

Scheme 19.   Structure of planar-chiral paracyclophanes, (*R*)-CP as a sensitizer and chirogenic host for photoisomerization of (*Z,Z*)-1,5-cyclooctadiene, **1,5-COD-ZZ** to **1,5-COD-EZ**.

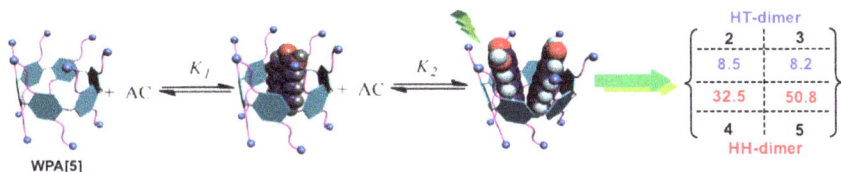

Scheme 20. Schematic illustration of stepwise 1:1 and 1:2 complexation of **WPA[6]** with AC for the photocyclodimerization and observed photodimer distributions. Reprinted with permission from Ref. [85]. Copyright 2016, ScienceDirect, Elsevier.

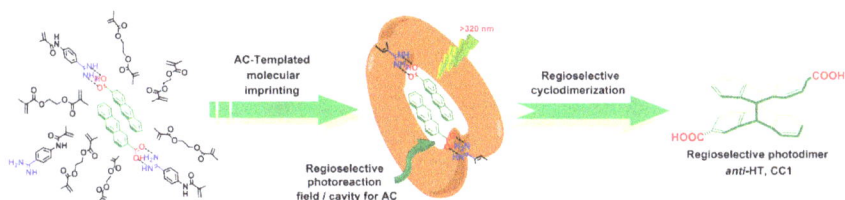

Scheme 21. Illustration of AC-template MIP reaction cavities created for regioselective photocyclodimerization of AC.

prepared and used as supramolecular photochirogenic hosts for the enantiodifferentiating photocyclodimerization of AC to improve the yield of the HH photodimers (HH/HT ratio up to 4.97) (Scheme 20).[85] In aqueous solution, WP[6] formed subsequently the 1:1 and 1:2 complexes (binding constants; $K_{1:1} = 1.21 \times 10^4$ M^{-1} and $K_{1:2} = 94$ M^{-1}) with AC, while the photoreaction afforded unfavorable HH photodimers, **CC3** and **CC4**, in a high yield of up to 83% with the HH/HT ratio being 4.97. The obtained enhanced yield of HH photodimer was due to the functionalized cationic groups in the periphery of **WPA[6]**, which improved the electrostatic attraction and simultaneously reduced the repulsion between the carboxylate anions of the HH stacked arrangement of AC in the cavity of **WPA[6]** in the ground state.[85]

Recently, Takeuchi *et al.*, reported the AC-templated cavities in the corresponding molecularly imprinted polymers (MIP) functioning as a supramolecular regioselective reaction field and a cavity for the photocyclodimerization of AC. This system afforded the *anti*-HT photodimer, **CC1**, exclusively among four other possible regioisomeric cyclodimers (Scheme 21).[86] The AC-imprinted cavities were constructed by the precipitation polymerization of both functional (*N*-methacryloyl-4-aminobenzamidine) and cross-linking (ethylene glycol dimethacrylate) monomers. The created MIP cavities preferentially bound to AC with high affinity and arranged in the *anti*-head-to-tail stacking manner that results in subsequent photoreaction, specifically yielding *anti*-HT, **CC1** as a regioselective photodimer.

So far, a wide variety of strategies and supramolecular chiral hosts/ assemblies like molecular clips, gelated-nanosponges, chirally coordinated transition metal ions, bio-supramolecular hosts such as proteins [plasma proteins — serum albumins (BSA and HSA), chaperone protein — prefoldin], antibodies, DNA,[87–91] and heterogeneous system such as chiral zeolites and mesoporous silica[7c,92–94] have been successfully employed for photochirogenic reactions. The systematic study of each photochirogenic reaction mediated in supramolecular systems significantly provides key tools to control and manipulate the reactivity, switching of the selectivity of photo-products, and the whole involved chirogenic process, by judicious choice of suitably modified host, solvent/media, pH, irradiation wavelength, and use of additives.

4. Concluding Remarks and Feature Perspectives

Chirogenesis is a unique process occurring in natural systems and an interdisciplinary emerging field of science in recent years. In this respect, a wide variety of strategies have been proposed; among them, supramolecular systems are the most fascinating, diverse, and adaptable. The major advantage of these tailored supramolecular architectures is their ability to encapsulate guest molecules *via* non-covalent interactions for effective transfer of asymmetric information with the aid of external stimuli (light, pH, media, and additives), resulting in the conformational changes in the whole supramolecular structure, rendering it as chiral.

Supramolecular chemistry is considered as one of the most sophisticated arts of science, being well-suited for designing and constructing various chirogenic hosts to mimic natural systems and processes. These supramolecular chirogenic systems (hosts and whole assemblies) are generally classified into three main categories, based on the nature and origin of chirality generated outcome (inherent, induced, and amplified), which can be obtained through individually selected methods and strategies.

These supramolecular chirogenic systems are of a high level of functional advancement, being effectively employed in various led application studies, in which chiral induction, sensing/recognition, discrimination, and photochirogenesis processes are of paramount importance. The chirogenesis efficiency can be finely tuned by judicious choice of suitably modified chirogenic host by screening its size and functionality, appropriate guest molecules as chiral/achiral substrates, nature of the medium/solvent used, and external stimuli. These systematic and comprehensive studies achieved exciting potential advances as in fundamental science and in various modern applications. However, there are still significant challenges in the prospective design and construction of radically novel adaptive supramolecular chirogenic systems possessing innovative properties, which can be developed for use in new practical technologies.

Particularly, further chirogenic perspectives and approaches in the modern interdisciplinary field of research on designing and creating various supramolecular chirogenic systems move toward the adaptive functionality in chiral sensing/recognition, discrimination, photochirogenesis, and other applications. Another prospective application can be the development of new hybrid supramolecular chirogenic systems, which can be used as novel sensory self-assemblies for various types of chirality ranging from the smallest chemical differences (such as differentiation of proton and deuterium) and up to large biological objects, such as DNA, proteins, or even whole cells.

Acknowledgments

We acknowledge the support of this work by the Estonian Research Council through Grant PRG399, the H2020-FETOPEN 828779 INITIO project, National Natural Science Foundation of China (No. 21871194, 21572142), National Key Research and Development Program of China (No. 2017YFA0505903), Science & Technology Department of Sichuan Province (2017SZ0021), and the start-up research grant (YZZ16005) from South-Central University for Nationalities.

References

1. R. S. Cahn, C. Ingold, and V. Prelog, Specification of molecular chirality, *Angew. Chem., Int. Ed. Engl.* **5**, 385–415 (1966).
2. (a) G. A. Hembury, V. V. Borovkov, and Y. Inoue, Chirality-Sensing supramolecular systems, *Chem. Rev.* **108**, 1–73 (2008); (b) M. Liu, L. Zhang, and T. Wang, Supramolecular chirality in self-assembled systems, *Chem. Rev.* **115**, 7304–7397 (2015); (c) A. Sumna, Inherently chiral concave molecules-from synthesis to applications, *Chem. Soc. Rev.* **39**, 4274–4285 (2010).
3. (a) S. M. Goldup, A chiral catalyst with a ring to it, *Nat. Chem.* **8**, 404–406 (2016); (b) R. Katoono, K. Kusaka, S. Kawai, Y. Tanaka, K. Hanada,

T. Nehira, K. Fujiwara, and T. Suzuki, Chiroptical molecular propellers based on hexakis(phenylethynyl)benzene through the complexation-induced intramolecular transmission of local point chirality, *Org. Biomol. Chem.* **12**, 9532–9538 (2014); (c) J.-C. Chambron and F. Richard Keene, "*Principles of molecular chirality*", in "*Chirality in Supramolecular Assemblies: Causes and Consequences*", First Edition, F. Richard Keene, (Eds.), John Wiley & Sons, Ltd., Hoboken, New Jersey, United States. pp. 1–43 (2017).

4. (a) J. R. Brandt, F. Salerno, and M. Fuchter, The added value of small-molecule chirality in technological applications, *Nat. Chem. Rev.* **1**, 0045 (2017); (b) W. Auwärter, D. Écija, F. Klappenberger, and J. V. Barth, Porphyrins at interfaces, *Nat. Chem.* **7**, 105–120 (2015).

5. (a) X. Ma and Y. Zhao, Biomedical applications of supramolecular systems based on host–guest interactions, *Chem. Rev.* **115**, 7794–7839 (2015); (b) X. Yan, F. Wang, B. Zheng, and F. Huang, Stimuli-responsive supramolecular polymeric materials, *Chem. Soc. Rev.* **41**, 6042–6065 (2012); (c) Y. Zhao, F. Sakai, L. Su, Y. Liu, K. Wei, G. Chen, and M. Jiang, Progressive macromolecular self — assembly: From biomimetic chemistry to bio-inspired materials, *Adv. Mat.* **25**, 5215–5256 (2013).

6. (a) M. V. Escarcega-Bobadilla and A. W. Kleij, Artificial chirogenesis: A gateway to new opportunities in material science and catalysis, *Chem. Sci.* **3**, 2421–2428 (2012); (b) C. Yang and Y. Inoue, Supramolecular photochirogenesis, *Chem. Soc. Rev.* **43**, 4123–4143 (2014); (c) C. Yang, Recent progress in supramolecular chiral photochemistry, *Chin. Chem. Lett.* **24**, 437–441 (2013); (d) Z. Yan, W. Wu, C. Cheng, and Y. Inoue, Catalytic supramolecular photochirogenesis, *Supramol. Catal.* **2**, 9–24 (2015); (e) V. V. Borovkov, G. A. Hembury, and Y. Inoue, Origin, control, and application of supramolecular chirogenesis in bisporphyrin-based systems, *Acc. Chem. Res.* **37**, 449–459 (2004); (f) M. Hasan and V. Borovkov, Helicene-based chiral auxiliaries and chirogenesis, *Symmetry* **10**, 10 (2018); (g) V. Borovkov, Effective supramolecular chirogenesis in ethane-bridged bis-porphyrinoids, *Symmetry* **2**, 184–200 (2010).

7. (a) V. Borovkov, Guest (Ed.), Special Issue *Chiral Auxiliaries and Chirogenesis*, *Symmetry* ISSN 2073–8994 (2017) (https://www.mdpi.com/journal/symmetry/special_issues/Chiral_Auxiliarie); (b) H. Lu and N. Kobayashi, Optically active porphyrin and phthalocyanine systems, *Chem. Rev.* **116**, 6184–6261 (2016); (c) C. Yang and Y. Inoue, in *Supramolecular Photochemistry: Controlling Photochemical Processes*, V. Ramamurthy and Y. Inoue (Eds.), John Wiley & Sons Inc., Hoboken, New Jersey, United States, p. 115 (2011).

8. D. Voet and J. G. Voet, *Biochemistry*, Second Edition, Wiley, New York, p. 1361 (1995), ISBN 0-471-58651-X.

9. E. Yashima, N. Ousaka, D. Taura, K. Shimomura, T. Ikai, and K. Maeda, Supramolecular helical systems: Helical assemblies of small molecules,

foldamers, and polymers with chiral amplification and their functions, *Chem. Rev.* **116**, 13752–13990 (2016).

10. (a) V. Ramamurthy and J. Sivaguru, Supramolecular photochemistry as a potential synthetic tool: Photocycloaddition, *Chem. Rev.* **116**, 9914–9993 (2016); (b) B. C. Pemberton, R. Raghunathan, S. Volla, and J. Sivaguru, From containers to catalysts: Supramolecular catalysis within cucurbiturils, *Chem. Eur-J.* **18**, 12178–12190 (2012); (c) P. Duan, H. Cao, L. Zhang, and M. Liu, Gelation induced supramolecular chirality: Chirality transfer, amplification and application, *Soft Matter* **10**, 5428–5448 (2014); (d) J. A. Kitchen and P. A. Gale, Complexity of supramolecular assemblies, in *Chirality in Supramolecular Assemblies: Causes and Consequences*, First Edition, F. Richard Keene (Ed.), John Wiley & Sons, Ltd., pp. 94–141 (2017).

11. K. Kanagaraj, K. Lin, W. Wu, G. Gao, Z. Zhong, D. Su, and C. Yang, Chiral buckybowl molecules, *Symmetry* **9**, 174 (2017).

12. (a) IUPAC. Compendium of Chemical Terminology, 2nd ed. (the "Gold Book"). Compiled by A. D. McNaught and A. Wilkinson. Blackwell Scientific Publications, Oxford (1997). Online version (2019) created by S. J. Chalk. ISBN 0-9678550-9-8; (b) G. Pescitelli, N. Sreerama, P. Salvadori, K. Nakanishi, N. Berova and R. W. Woody, Inherent chirality dominates the visible/near-ultraviolet CD spectrum of rhodopsin, *J. Am. Chem. Soc.* **130**, 6170-6181 (2008), and references therein.

13. P. Xing and Y. Zhao, Controlling supramolecular chirality in multi-component self-assembled systems, *Acc. Chem. Res.* **51**, 2324–2334 (2018).

14. (a) D. B. Amabilino and J. Veciana, *Supramolecular Chiral Functional Materials*, Springer, Berlin, Heidelberg, p. 253 (2006); (b) Y. Wang, J. Xu, Y. Wang, and H. Chen, Emerging chirality in nanoscience, *Chem. Soc. Rev.* **42**, 2930–2962 (2013); (c) S. De Feyter and F. C. De Schryver, Two-dimensional supramolecular self-assembly probed by scanning tunneling microscopy, *Chem. Soc. Rev.* **32**, 139–150 (2003).

15. (a) J. A. A. W. Elemans, I. De Cat, H. Xu, and S. De Feyter, Two-dimensional chirality at liquid–solid interfaces, *Chem. Soc. Rev.* **38**, 722–736 (2009); (b) D. Y. Du, L. K. Yan, Z. M. Su, S. L. Li, Y. Q. Lan, and E. B. Wang, Chiral polyoxometalate-based materials: From design syntheses to functional applications, *Coord. Chem. Rev.* **257**, 702–717 (2013); (c) J. Crassous, Chiral transfer in coordination complexes: Towards molecular materials, *Chem. Soc. Rev.* **38**, 830–840 (2009); (d) E. C. Constable, Stereogenic metal centres — from Werner to supramolecular chemistry, *Chem. Soc. Rev.* **42**, 1637–1651 (2013); (e) V. K. Valev, J. J. Baumberg, C. Sibilia, and T. Verbiest, Chirality and chiroptical effects in plasmonic nanostructures: Fundamentals, recent progress, and outlook, *Adv. Mater.* **25**, 2517–2534 (2013); (f) K. Nonaka,

M. Yamaguchi, and M. Yasui, Guest-induced supramolecular chirality in a ditopic azoprobe–cyclodextrin complex in water, *Chem. Commun.* **26**, 10059–10061 (2014).

16. S. Lifson, C. Andreola, N. C. Peterson, and M. M. Green, Macromolecular stereochemistry: Helical sense preference in optically active polyisocyanates. Amplification of a conformational equilibrium deuterium isotope effect, *J. Am. Chem. Soc.* **111**, 8850–8858 (1989).

17. (a) A. R. A. Palmans, J. Vekemans, R. A. Hikmet, H. Fischer, and E. W. Meijer, Lyotropic liquid-crystalline behavior in disc-shaped compounds incorporating the 3,3′-di(acylamino)-2,2′-bipyridine unit, *Adv. Mater.* **10**, 873–876 (1998); (b) I. Danila, F. Riobe, F. Piron, J. Puigmarti-Luis, J. D. Wallis, M. Linares, H. Ågren, D. Beljonne, D. B. Amabilino, and N. C. Avarvari, Hierarchical chiral expression from the nano- to mesoscale in synthetic supramolecular helical fibers of a monamphiphilic C3-symmetrical π-functional molecule, *J. Am. Chem. Soc.* **133**, 8344–8353 (2011); (c) A. R. A. Palmans, J. Vekemans, E. E. Havinga, and E. W. Meijer, Sergeants-and-soldiers principle in chiral columnar stacks of disc-shaped molecules with C3 symmetry, *Angew. Chem., Int. Ed.* **36**, 2648–2651 (1997); (d) A. R. Palmans and E. W. Meijer, Amplification of chirality in dynamic supramolecular aggregates, *Angew. Chem., Int. Ed.* **46**, 8948–8968 (2007); (e) C. Kulkarni, R. Munirathinam, and S. J. George, Self-assembly of coronene bisimides: Mechanistic insight and chiral amplification, *Chem.-Eur. J.* **19**, 11270–11278 (2013); (f) M. M. J. Smulders, P. J. M. States, T. Mes, T. F. E. Paffen, A. P. H. J. Schenning, A. R. A. Palmans, and E. W. Meijer, Probing the limits of the majority-rules principle in a dynamic supramolecular polymer, *J. Am. Chem. Soc.* **132**, 620–626 (2010); (g) J. van Gestel, P. van der Schoot, and M. A. J. Michels, Amplification of chirality in helical supramolecular polymers beyond the long-chain limit, *J. Chem. Phys.* **120**, 8253–8261 (2004).

18. (a) B. Nie, T. Zhan, T. Zhou, Z. Xiao, G. Jiang, and X. Zhao, Self-assembly of chiral propeller-like supermolecules with unusual "Sergeants-and-Soldiers" and "Majority-Rules" effects, *Chem.-Asian J.* **9**, 754–758 (2014); (b) T. Seki, A. Asano, S. Seki, Y. Kikkawa, H. Murayama, T. Karatsu, A. Kitamura, and S. Yagai, Rational construction of perylene bisimide columnar superstructures with a biased helical sense, *Chem.-Eur. J.* **17**, 3598–3608 (2011).

19. (a) F. Aparicio, F. Vicente, and L. Sanchez, Amplification of chirality in N,N′-1,2-ethanediylbisbenzamides: From planar sheets to twisted ribbons, *Chem. Commun.* **46**, 8356–8358 (2010); (b) H. Cao, X. Zhu, and M. Liu, Self-assembly of racemic alanine derivatives: Unexpected chiral twist and enhanced capacity for the discrimination of chiral species, *Angew. Chem., Int. Ed.* **52**, 4122–4126 (2013).

20. (a) E. Yashima, K. Maeda, H. Iida, Y. Furusho, and K. Nagai, Helical polymers: Synthesis, structures, and functions, *Chem. Rev.* **109**, 6102–6211 (2009); (b) S. Takashima, H. Abe, and M. Inouye, Copper(ii)/phenanthroline-mediated CD-enhancement and chiral memory effect on a meta-ethynylpyridine oligomer, *Chem. Commun.* **48**, 3330–3332 (2012); (c) E. Bellacchio, R. Lauceri, S. Gurrieri, L. M. Scolaro, A. Romeo, and R. Purrello, Template-imprinted chiral porphyrin aggregates, *J. Am. Chem. Soc.* **120**, 12353–12354 (1998); (d) R. Purrello, A. Raudino, L. M. Scolaro, A. Lois, E. Bellacchio, and R. Lauceri, Ternary porphyrin aggregates and their chiral memory, *J. Phys. Chem. B* **104**, 10900–10908 (2000); (e) L. Rosaria, A. D' Urso, A. Mammana, and R. Purrello, Chiral memory: Induction, amplification, and switching in porphyrin assemblies, *Chirality* **20**, 411–419 (2008); (f) A. Mammana, A. D' Urso, R. Lauceri, and R. Purrello, Switching off and on the supramolecular chiral memory in porphyrin assemblies, *J. Am. Chem. Soc.* **129**, 8062–8063 (2007); (g) L. Zhao, R. Xiang, R. Ma, X. Wang, Y. An, and L. Shi, Chiral conversion and memory of TPPS J-aggregates in complex micelles: PEG-b-PDMAEMA/TPPS, *Langmuir* **27**, 11554–11559 (2011); (h) F. Helmich, M. M. Smulders, C. C. Lee, A. P. Schenning, and E. W. Meijer, Effect of stereogenic centers on the self-sorting, depolymerization, and atropisomerization kinetics of porphyrin-based aggregates, *J. Am. Chem. Soc.* **133**, 12238–12246 (2011); (i) F. Helmich, C. C. Lee, A. P. H. J. Schenning, and E. W. Meijer, Chiral memory via chiral amplification and selective depolymerization of porphyrin aggregates, *J. Am. Chem. Soc.* **132**, 16753–16755 (2010); (j) J. Zhao, Y. Ruan, R. Zhou, and Y. Jiang, Memory of chirality in J-type aggregates of an achiral perylene dianhydride dye created in a chiral asymmetric catalytic synthesis, *Chem. Sci.* **2**, 937–944 (2011); (k) W. Zhang, W. Jin, T. Fukushima, N. Ishii, and T. Aida, Dynamic or nondynamic? Helical trajectory in hexabenzocoronene nanotubes biased by a detachable chiral auxiliary, *J. Am. Chem. Soc.* **135**, 114–117 (2013); (l) A. M. Castilla, N. Osuka, R. A. Bilbeisi, E. Valeri, T. K. Ronson, and J. R. Nitschke, High-fidelity stereochemical memory in a FeII4L4 tetrahedral capsule, *J. Am. Chem. Soc.* **135**, 17999–18006 (2013).
21. (a) A. J. Savyasachi, O. Kotova, S. Shanmugaraju, S. J. Bradberry, G. M. O'Maille, and T. Gunnlaugsson, Supramolecular chemistry: A toolkit for soft functional materials and organic particles, *Chem.* **3**, 764–811 (2017); (b) J. Shen and Y. Okamoto, Efficient separation of enantiomers using stereoregular chiral polymers, *Chem. Rev.* **116**, 1094–1138 (2016).
22. K. Ariga, G. J. Richards, S. Ishihara, H. Izawa, and J. P. Hill, Intelligent chiral sensing based on supramolecular and interfacial concepts, *Sensors* **10**, 6796–6820 (2010).
23. J. L. Serrano and T. Sierra, Helical supramolecular organizations from metal-organic liquid crystals, *Coord. Chem. Rev.* **242**, 73–85 (2003).

24. H. W. Brooks, C. W. Guida, and G. K. Daniel, The significance of chirality in drug design and development, *Curr. Top. Med. Chem.* **11**, 760–770 (2011).
25. J. Leffingwell and D. Leffingwell, *Spec. Chem. Mag.* **30** (2011).
26. (a) F. Bellina, D. L. Mendola, C. Pedone, E. Rizzarelli, M. Savianoc, and G. Vecchio, Selectively functionalized cyclodextrins and their metal complexes, *Chem. Soc. Rev.* **38**, 2756–2781 (2009); (b) A. Concheiro and C. Alvarez-Lorenzo, Chemically crosslinked and grafted cyclodextrin hydrogels: From nanostructures to drug-eluting medical devices, *Adv. Drug Deliv. Rev.* **65**, 1188–1203 (2013); (c) W.-F. Lai, A. L. Rogach, and W.-T. Wong, Chemistry and engineering of cyclodextrins for molecular imaging, *Chem. Soc. Rev.* **46**, 6379–6419 (2017).
27. (a) M. Blanco and I. Valverde, Choice of chiral selector for enantiosepara-tion by capillary electrophoresis, *Trends Anal. Chem.* **22**, 428–439 (2003); (b) T. Cserhati, New applications of cyclodextrins in electrically driven chromatographic systems: A review, *Biomed. Chromatogr.* **22**, 563–571 (2008); (c) K. Takahashi, Organic reactions mediated by cyclodextrins, *Chem. Rev.* **98**, 2013–2034 (1998); (d) F. Cramer and W. Dietsche, Über Einschlußverbindungen, XV. Spaltung von racematen mit cyclodextrinen, *Chem. Ber.* **92**, 378–384 (1959).
28. (a) M. Rekharsky and Y. Inoue, Chiral recognition thermodynamics of β-cyclodextrin: The thermodynamic origin of enantioselectivity and the enthalpy–entropy compensation effect, *J. Am. Chem. Soc.* **122**, 4418–4435 (2000); (b) M. Rekharsky and Y. Inoue, Complexation thermodynamics of cyclodextrins, *J. Am. Chem. Soc.* **122**, 4418–4435 (2000); (c) M. V. Rekharsky and Y. Inoue, Complexation thermodynamics of γ-cyclodextrin with N-carbobenzyloxy aromatic amino acids and ω-phenylalkanoic acids, *Chem. Rev.* **98**, 1875–1918 (1998); (d) M. Rekharsky and Y. Inoue, *J. Am. Chem. Soc.* **122**, 10949–10955 (2000).
29. M. V. Rekharsky and Y. Inoue, Complexation and chiral recognition thermo-dynamics of 6-amino-6-deoxy-β-cyclodextrin with anionic, cationic, and neutral chiral guests: Counterbalance between van der waals and coulombic interactions, *J. Am. Chem. Soc.* **124**, 813–826 (2002).
30. (a) H. Yamamura, M. V. Rekharsky, Y. Ishihara, M. Kawai, and Y. Inoue, Factors controlling the complex architecture of native and modified cyclo-dextrins with dipeptide (Z-Glu-Tyr) studied by microcalorimetry and NMR spectroscopy: Critical effects of peripheral bis-trimethylamination and cavity size, *J. Am. Chem. Soc.* **126**, 14224–14233 (2004); (b) T. Ogoshi and A. Harada, Chemical sensors based on cyclodextrin derivatives, *Sensors* **8**, 4961–4982 (2008).
31. L. Dai, W. Wu, W. Liang, W. Chen, X. Yu, J. Ji, C. Xiao, and C. Yang, Enhanced chiral recognition by γ-cyclodextrin–cucurbit[6]uril-cowheeled [4]pseudorotaxanes, *Chem. Commun.* **54**, 2643–2646 (2018).

32. Q. Huang, L. Jiang, W. Liang, J. Gui, D. Xu, W. Wu, Y. Nakai, M. Nishijima, G. Fukuhara, T. Mori, Y. Inoue, and C. Yang. Inherently chiral azonia[6] helicene-modified β-cyclodextrin: Synthesis, characterization, and chirality sensing of underivatized amino acids in water, *J. Org. Chem.* **81**, 3430–3434 (2016).

33. (a) N. Kobayashi, Optically active phthalocyanines, *Coord. Chem. Rev.* **219–221**, 99–123 (2001); (b) N. Kobayashi, Optically active porphyrin systems analyzed by circular dichroism, in *Handbook of Porphyrin Science*, Vol. 7, Ch. 33, K. Kadish, K. Smith, and R. Guilard (Eds.), World Scientific, Singapore, pp. 147–240 (2010).

34. J. L. Sessler, Porphyrin analogues, *J. Porphyrins Phthalocyanines* **4**, 331–336 (2000).

35. (a) V. Valderrey, G. Aragay, and P. Ballester, Porphyrin tweezer receptors: Binding studies, conformational properties and applications, *Coord. Chem. Rev.* **258–259**, 137–156 (2014); (b) V. Borovkov, Supramolecular Chirality in Porphyrin Chemistry, *Symmetry* **6**, 256–294 (2014).

36. (a) H. Ogoshi and T. Mizutani, Multifunctional and chiral porphyrins: Model receptors for chiral recognition, *Acc. Chem. Res.* **31**, 81–89 (1998); (b) J.-C. Marchon and R. Ramasseul, Phthalocyanines: Spectroscopic and electrochemical characterization, in *The Porphyrin Handbook*, K. M. K. M. S. Guilard (Ed.), Academic Press, Amsterdam, (2003), ISBN 978-0-08-092390-1, https://doi.org/10.1016/C2009-0-22719-X.

37. (a) V. V. Borovkov, N. Zh Mamardashvili, and Y. Inoue, Optically active supramolecular systems based on porphyrins, *Russ. Chem. Rev.* **75**, 737–748 (2006); (b) G. Simonneaux and P. Le Maux, Optically active ruthenium porphyrins: Chiral recognition and asymmetric catalysis, *Coord. Chem. Rev.* **228**, 43–60 (2002); (c) E. Rose, B. Andrioletti, S. Zrig, and M. Quelquejeu-Etheve, Enantioselective epoxidation of olefins with chiral metalloporphyrincatalysts, *Chem. Soc. Rev.* **34**, 573–583 (2005).

38. (a) N. Harada and K. Nakanishi, *Circular Dichroic Spectroscopy: Exciton Coupling in Organic Stereochemistry,* University Science Books Mill Valley, CA (1983); (b) N. Berova, L. D. Bari, and G. Pescitelli, Application of electronic circular dichroism in configurational and conformational analysis of organic compounds, *Chem. Soc. Rev.* **36**, 914–931 (2007).

39. Y. Ding, W.-H. Zhu, and Y. Xie, Development of ion chemosensors based on porphyrin analogues, *Chem. Rev.* **117**, 2203–2256 (2017).

40. (a) V. V. Borovkov, J. M. Lintuluoto, M. Fujiki, and Y. Inoue, Temperature effect on supramolecular chirality induction in bis(zinc porphyrin), *J. Am. Chem. Soc.* **122**, 4403–4407 (2000); (b) V. V. Borovkov, J. M. Lintuluoto, and Y. Inoue, Supramolecular chirogenesis in zinc porphyrins: Mechanism, role of guest structure, and application for the absolute configuration determination, *J. Am. Chem. Soc.* **123**, 2979–2989 (2001); (c) V. V. Borovkov,

J. M. Lintuluoto, and Y. Inoue, Stoichiometry-controlled supramolecular chirality induction and inversion in bisporphyrin systems, *Org. Lett.* **4**, 169–171 (2002); (d) V. V. Borovkov, J. M. Lintuluoto, H. Sugeta, M. Fujiki, R. Arakawa, and Y. Inoue, Supramolecular chirogenesis in zinc porphyrins: Equilibria, binding properties, and thermodynamics, *J. Am. Chem. Soc.* **124**, 2993–3006 (2002); (e) V. V. Borovkov, J. M. Lintuluoto, M. Sugiura, and Y. Inoue, Remarkable stability and enhanced optical activity of a chiral supramolecular bis-porphyrin tweezer in both solution and solid state, *J. Am. Chem. Soc.* **124**, 11282–11283 (2002); (f) J. M. Lintuluoto, V. V. Borovkov, and Y. Inoue, Direct determination of absolute configuration of monoalcohols by bis(magnesium porphyrin), *J. Am. Chem. Soc.* **124**, 13676–13677 (2002).

41. (a) V. V. Borovkov, G. A. Hembury, and Y. Inoue, The origin of solvent-controlled supramolecular chirality switching in a bis(zinc porphyrin) system, *Angew. Chem.* **115**, 5468–5472 (2003); (b) V. V. Borovkov, J. M. Lintuluoto, G. A. Hembury, M. Sugiura, R. Arakawa, and Y. Inoue, Supramolecular chirogenesis in zinc porphyrins: Interaction with bidentate ligands, formation of tweezer structures, and the origin of enhanced optical activity, *J. Org. Chem.* **68**, 7176–7192 (2003); (c) V. V. Borovkov, I. Fujii, A. Muranaka, G. A. Hembury, T. Tanaka, A. Ceulemans, N. Kobayashi, and Y. Inoue, Rationalization of supramolecular chirality in a bisporphyrin system, *Angew. Chem. Int. Ed.* **43**, 5481–5485 (2004); (d) V. V. Borovkov, G. A. Hembury, and Y. Inoue, Supramolecular chirogenesis with bischlorin versus bis-porphyrin hosts: Peculiarities of chirality induction and modulation of optical activity, *J. Org. Chem.* **70**, 8743–8754 (2005); (e) V. V. Borovkov, A. Muranaka, G. A. Hembury, Y. Origane, G. V. Ponomarev, N. Kobayashi, and Y. Inoue, Chiral bis-chlorin: Enantiomer resolution and absolute configuration determination, *Org. Lett.* **7**, 1015–1018 (2005).

42. (a) V. V. Borovkov and Y. Inoue, Supramolecular chiral recognition by bischlorins: A two-point interaction mode combined with the host's conformational modulation controlled by the guest's stereochemistry and bulkiness, *Org. Lett.* **8**, 2337–2340 (2006); (b) P. Bhyrappa, V. V. Borovkov, and Y. Inoue, Supramolecular chirogenesis in bis-porphyrins: Interaction with chiral acids and application for the absolute configuration assignment, *Org. Lett.* **9**, 433–435 (2007); (c) V. V. Borovkov, J. M. Lintuluoto, and Y. Inoue, Syn-anti conformational changes in zinc porphyrin dimers induced by temperature-controlled alcohol ligation, *J. Phys. Chem. B* **103**, 5151–5156 (1999); (d) V. V. Borovkov, T. Harada, Y. Inoue, and R. Kuroda, Phase — sensitive supramolecular chirogenesis in bisporphyrin systems, *Angew. Chem. Int. Ed.* **41**, 1378–1381 (2002).

43. (a) T. Hayashi, M. Nonoguchi, T. Aya, and H. Ogoshi, Molecular recognition of α,ω-diamines by metalloporphyrin dimer, *Tetrahedron Lett.* **38**, 1603–1606 (1997); (b) T. Ema, N. Ouchi, T. Doi, T. Korenaga, and T. Sakai, Highly sensitive chiral shift reagent bearing two zinc porphyrins, *Org. Lett.* **7**, 3985–3988 (2005).

44. M. Anyika, H. Gholami, K. D. Ashtekar, R. Acho, and B. Borhan, Point-to-axial chirality transfer—a new probe for "sensing" the absolute configurations of monoamines, *J. Am. Chem. Soc.* **136**, 550–553 (2014).

45. (a) S. Brahma, S. A. Ikbal, and S. P. Rath, Synthesis, Structure, and properties of a series of chiral tweezer–diamine complexes consisting of an achiral zinc(II) bisporphyrin host and chiral diamine guest: Induction and rationalization of supramolecular chirality, *Inorg. Chem.* **53**, 49–62 (2014); (b) S. A. Ikbal, S. Brahma, and S. P. Rath, Transfer and control of molecular chirality in the 1:2 host–guest supramolecular complex consisting of Mg(ii)bisporphyrin and chiral diols: The effect of H-bonding on the rationalization of chirality, *Chem. Commun.* **50**, 14037–14040 (2014).

46. (a) H. Wynberg, Some observations on the chemical, photochemical, and spectral properties of thiophenes, *Acc. Chem. Res.* **4**, 65–73 (1971); (b) Y. Shen and C.-F. Chen, helicenes: Synthesis and applications, *Chem. Rev.* **112**, 1463–1535 (2012).

47. (a) M. Gingras, One hundred years of helicene chemistry. Part 3: Applications and properties of carbohelicenes, *Chem. Soc. Rev.* **42**, 1051–1095 (2013); (b) K.-H. Ernst, Stereochemical recognition of helicenes on metal surfaces, *Acc. Chem. Res.* **49**, 1182–1190 (2016).

48. (a) H. Rau and F. Totter, Exciplex and ion pair quenching in a chiral hexaheliceneamine system, *J. Photochem. Photobiol. A* **63**, 337–347 (1992); (b) M. T. Reetz and S. Sostmann, 2,15-Dihydroxy-hexahelicene (HELIXOL): Synthesis and use as an enantioselective fluorescent sensor, *Tetrahedron* **57**, 2515–2520 (2001).

49. (a) M. Nakazaki, K. Yamamoto, T. Ikeda, T. Kitsuki, and Y. Okamoto, Synthesis and chiral recognition of novel crown ethers incorporating helicene chiral centres, *J. Chem. Soc. Chem. Commun.* 787–788 (1983); (b) K. Yamamoto, T. Ikeda, T. Kitsuki, Y. Okamoto, H. Chikamatsu, and M. Nakazaki, Synthesis and chiral recognition of optically active crown ethers incorporating a helicene moiety as the chiral centre, *J. Chem. Soc. Perkin Trans. 1*, 271–276 (1990).

50. (a) D. J. Weix, S. D. Dreher, and T. J. Katz, [5]HELOL Phosphite: A helically grooved sensor of remote chirality, *J. Am. Chem. Soc.* **122**, 10027–10032 (2000); (b) D. Z. Wang and T. J. Katz, A [5]HELOL analogue that senses remote chirality in alcohols, phenols, amines, and carboxylic acids, *J. Org. Chem.* **70**, 8497–8502 (2005).

51. X. Chen, Z. Huang, S.-Y. Chen, K. Li, X.-Q. Yu, and L. Pu, Enantioselective gel collapsing: A new means of visual chiral sensing, *J. Am. Chem. Soc.* **132**, 7297–7299 (2010).
52. T. Tu, W. Fang, X. Bao, X. Li, and K. H. Dtz, Visual chiral recognition through enantioselective metallogel collapsing: Synthesis, characterization, and application of platinum–steroid low-molecular-mass gelators, *Angew. Chem. Int. Ed.* **50**, 6601–6605 (2011).
53. (a) H. Jintoku, T. Sagawa, T. Sawada, M. Takafuji, and H. Ihara, Versatile chiroptics of peptide-induced assemblies of metalloporphyrins, *Org. Biomol. Chem.* **8**, 1344–1350 (2010); (b) H. Jintoku, M. Takafuji, R. Oda, and H. Ihara, Enantioselective recognition by a highly ordered porphyrin-assembly on a chiral molecular gel, *Chem. Commun.* **48**, 4881–4883 (2012).
54. (a) Y. Inoue and V. Ramamurthy (Eds.), *Molecular and Supramolecular Photochemistry* in *Chiral Photochemistry*, Volume 11, Marcel Dekker, New York, pp. 1–686 (2004) (ISBN 0-8247-5710-6); (b) C. Yang and Y. Inoue, Photochirogenesis, in *CRC Handbook of Organic Photochemistry and Photobiology*, Vol. 1, Third Edition, A. G. Griesbeck, M. Oelgemoller, and F. Ghetti (Eds.), Taylor & Francis Group, Boca Raton, p. 125 (2012).
55. V. Ramamurthy and S. Gupta, Supramolecular photochemistry: From molecular crystals to water-soluble capsules, *Chem. Soc. Rev.* **44**, 119–135 (2015).
56. (a) P. Suresh and K. Pitchumani, Per-6-amino-β-cyclodextrin catalyzed asymmetric Michael addition of nitromethane and thiols to chalcones in water, *Tetrahedron Asymmetry* **19**, 2037–2044 (2008); (b) K. Kanagaraj and K. Pitchumani, Per-6-amino-β-cyclodextrin as a chiral base catalyst promoting one-pot asymmetric synthesis of 2-Aryl-2,3-dihydro-4-quinolones, *J. Org. Chem.* **78**, 744–751 (2013); (c) K. Kanagaraj, P. Suresh, and K. Pitchumani, Per-6-amino-β-cyclodextrin as a reusable promoter and chiral host for enantioselective henry reaction, *Org. Lett.* **12**, 4070–4073 (2010); (d) C. C. Bai, B. R. Tian, T. Zhao, Q. Huang, and Z. Z. Wang, Cyclodextrin-catalyzed organic synthesis: Reactions, mechanisms, and applications, *Molecules* **22**, 1475 (2017); (e) G. Floresta, C. Talotta, C. Gaeta, M. D. Rosa, U. Chiacchio, P. Neri, and A. Rescifina, γ-Cyclodextrin as a catalyst for the synthesis of 2-methyl-3,5-diarylisoxazolidines in water, *J. Org. Chem.* **82**, 4631–4639 (2017).
57. (a) Y. Gao, M. Inoue, T. Wada, and Y. Inoue, Supramolecular photochirogenesis. 3. Enantiodifferentiating photoisomerization of cyclooctene included and sensitized by 6-O-mono(o-methoxybenzoyl)cyclodextrin, *J. Incl. Phenom. Macrocyclic Chem.* **50**, 111–118 (2004); (b) Y. Inoue, T. Wada, N. Sugahara, K. Yamamoto, K. Kimura, L.-H. Tong, X.-M. Gao, Z.-J. Hou, and Y. Liu, Supramolecular photochirogenesis. 2. Enantiodifferentiating photoisomerization of cyclooctene included and sensitized by 6-O-modified cyclodextrins, *J. Org. Chem.* **65**, 8041–8050 (2000); (c) C. Yang, T. Mori, T. Wada,

and Y. Inoue, Supramolecular enantiodifferentiating photoisomerization of (Z,Z)-1,3-cyclooctadiene included and sensitized by naphthalene-modified cyclodextrins, *New J. Chem.* **31**, 697–702 (2007).

58. (a) R. Lu, C. Yang, Y. Cao, Z. Wang, T. Wada, W. Jiao, T. Mori, and Y. Inoue, Enantiodifferentiating photoisomerization of cyclooctene included and sensitized by aroyl-β-cyclodextrins: A critical enantioselectivity control by substituents, *J. Org. Chem.* **73**, 7695–7701 (2008); (b) R. Lu, C. Yang, Y. Cao, Z. Wang, T. Wada, W. Jiao, T. Mori, and Y. Inoue, Supramolecular enantiodifferentiating photoisomerization of cyclooctene with modified β-cyclodextrins: Critical control by a host structure, *Chem. Commun.* 374–376 (2008); (c) G. Fukuhara, T. Mori, T. Wada, and Y. Inoue, Entropy-controlled supramolecular photochirogenesis: Enantiodifferentiating Z–E photoisomerization of cyclooctene included and sensitized by permethylated 6-O modified β-cyclodextrins, *J. Org. Chem.* **71**, 8233–8243 (2006); (d) G. Fukuhara, T. Mori, T. Wada, and Y. Inoue, Entropy-controlled supramolecular photochirogenesis: Enantiodifferentiating Z-E photoisomerization of cyclooctene included and sensitized by permethylated 6-O-benzoyl-b-cyclodextrin, *Chem. Commun.* 4199–4200 (2005).

59. (a) Y. Inoue, S. Kosaka, K. Matsumoto, H. Tsuneishi, T. Hakushi, A. Tai, K. Nakagawa, and L. Tong, Vacuum UV photochemistry in cyclodextrin cavities. Solid state Z-E photoisomerization of a cyclooctene-β-cyclodextrin inclusion complex, *J. Photochem. Photobiol. A* **71**, 61–64 (1993); (b) Y. Inoue, F. Dong, K. Yamamoto, L.-H. Tong, H. Tsuneishi, T. Hakushi, and A. Tai, Inclusion-enhanced optical yield and E/Z ratio in enantiodifferentiating photoisomerization of cyclooctene included and sensitized by β-cyclodextrin monobenzoate, *J. Am. Chem. Soc.* **117**, 11033–11034 (1995).

60. (a) W. Liang, C. Yang, M. Nishijima, G. Fukuhara, T. Mori, A. Mele, F. Castiglione, F. Caldera, F. Trotta, and Y. Inoue, Cyclodextrin nanosponge-sensitized enantiodifferentiating photoisomerization of cyclooctene and 1,3-cyclooctadiene, *Beilstein J. Org. Chem.* **8**, 1305–1311 (2012); (b) W. Liang, C. Yang, D. Zhou, H. Haneoka, M. Nishijima, G. Fukuhara, T. Mori, F. Castiglione, A. Mele, F. Caldera, F. Trotta, and Y. Inoue, Phase-controlled supramolecular photochirogenesis in cyclodextrin nanosponges, *Chem. Commun.* **49**, 3510–3512 (2013); (c) X. Wei, W. Liang, W. Wu, C. Yang, F. Trotta, F. Caldera, A. Mele, T. Nishimoto, and Y. Inoue, Solvent- and phase-controlled photochirogenesis. Enantiodifferentiating photoisomerization of (Z)-cyclooctene sensitized by cyclic nigerosylnigerose-based nanosponges crosslinked by pyromellitate, *Org. Biomol. Chem.* **13**, 2905–2912 (2015); (d) W. Liang, M. Zhao, X. Wei, Z. Yan, W. Wu, F. Caldera, F. Trotta, Y. Inoue, D. Su, Z. Zhong, and C. Yang, Photochirogenic nanosponges: Phase-controlled enantiodifferentiating photoisomerization of (Z)-cyclooctene

sensitized by pyromellitate-crosslinked linear maltodextrin, *RSC Adv.* **7**, 17184–17192 (2017).

61. A. Dawn, T. Shiraki, S. Haraguchi, H. Sato, K. Sada, and S. Shinkai, Transcription of chirality in the organogel systems dictates the enantiodifferentiating photodimerization of substituted anthracene, *Chem.-Eur. J.* **16**, 3676–3689 (2010).

62. Y. Ishida, Y. Kai, S.-Y. Kato, A. Misawa, S. Amano, Y. Matsuoka, and K. Saigo, Two-component liquid crystals as chiral reaction media: Highly enantioselective photodimerization of an anthracene derivative driven by the ordered microenvironment, *Angew. Chem., Int. Ed.* **47**, 8241–8245 (2008).

63. Z. Yan, Q. Huang, W. Liang, X. Yu, D. Zhou, W. Wu, J. J. Chruma, and C. Yang, Enantiodifferentiation in the photoisomerization of (Z,Z)-1,3-cyclooctadiene in the cavity of γ-cyclodextrin–curcubit[6]uril-wheeled [4]rotaxanes with an encapsulated photosensitizer, *Org. Lett.* **19**, 898–901 (2017).

64. (a) T. Tamaki and T. Kokubu, Acceleration of the photodimerization of water-soluble anthracenes included by β- and γ-cyclodextrins, *J. Incl. Phenom. Macrocyclic Chem.* **2**, 815–822 (1984); (b) T. Tamaki, T. Kokubu, and K. Ichimura, Regio- and stereoselective photodimerization of anthracene derivatives included by cyclodextrins, *Tetrahedron* **43**, 1485–1494 (1987); (c) A. Nakamura and Y. Inoue, Supramolecular catalysis of the enantiodifferentiating [4 + 4] photocyclodimerization of 2-anthracenecarboxylate by γ-cyclodextrin, *J. Am. Chem. Soc.* **125**, 966–972 (2003).

65. X. Wei, W. Wu, R. Matsushita, Z. Yan, D. Zhou, J. J. Chruma, M. Nishijima, G. Fukuhara, T. Mori, Y. Inoue, and C. Yang, Supramolecular photochirogenesis driven by higher-order complexation: Enantiodifferentiating photocyclodimerization of 2-anthracenecarboxylate to slipped cyclodimers via a 2:2 complex with β-cyclodextrin, *J. Am. Chem. Soc.* **140**, 3959–3974 (2018).

66. (a) C. Yang, A. Nakamura, G. Fukuhara, Y. Origane, T. Mori, T. Wada, and Y. Inoue, Pressure and temperature-controlled enantiodifferentiating [4+4] photocyclodimerization of 2-anthracenecarboxylate mediated by secondary faceand skeleton-modified γ-cyclodextrins, *J. Org. Chem.* **71**, 3126–3136 (2006); (b) H. Ikeda, T. Nihei, and A. Ueno, *J. Org. Chem.* **70**, 1237–1242 (2005); (c) A. Nakamura and Y. Inoue, Electrostatic manipulation of enantiodifferentiating photocyclodimerization of 2-anthracenecarboxylate within γ-cyclodextrin cavity through chemical modification. Inverted product distribution and enhanced enantioselectivity, *J. Am. Chem. Soc.* **127**, 5338–5339 (2005).

67. (a) Y. Gao, T. Wada, K. Yang, K. Kim, and Y. Inoue, Supramolecular photochirogenesis in sensitizing chiral nanopore: Enantiodifferentiating photoisomerization of (Z)-cyclooctene included and sensitized by POST-1, *Chirality* **17**,

S19–S23 (2005); (b) C. Yang, T. Mori, Y. Origane, Y. H. Ko, N. Selvapalam, K. Kim, and Y. Inoue, Highly stereoselective photocyclodimerization of α-cyclodextrin-appended anthracene mediated by γ-cyclodextrin and cucurbit[8]uril: A dramatic steric effect operating outside the binding site, *J. Am. Chem. Soc.* **130**, 8574–8575 (2008); (c) C. Yang, C. Ke, F. Kahee, D.-Q. Yuan, T. Mori, and Y. Inoue, pH-controlled supramolecular enantiodifferentiating photocyclodimerization of 2-anthracenecarboxylate with capped γ-cyclodextrins, *Aust. J. Chem.* **61**, 565–568 (2008).

68. J. Yao, Z. Yan, J. Ji, W. Wu, C. Yang, M. Nishijima, G. Fukuhara, T. Mori, and Y. Inoue, Ammonia-driven chirality inversion and enhancement in enantiodifferentiating photocyclodimerization of 2-anthracenecarboxylate mediated by diguanidino-γ-cyclodextrin, *J. Am. Chem. Soc.* **136**, 6916–6919 (2014).

69. J. Yi, W. Liang, X. Wei, J. Yao, Z. Yan, D. Su, Z. Zhong, G. Gao, W. Wu, and C. Yang, Switched enantioselectivity by solvent components and temperature in photocyclodimerization of 2-anthracenecarboxylate with 6A,6X-diguanidio–γ-cyclodextrins, *Chin. Chem. Lett.* **29**, 87–90 (2018).

70. (a) C. Yang, Q. Wang, M. Yamauchi, J. Yao, D. Zhou, M. Nishijima, G. Fukuhara, T. Mori, Y. Liu, and Y. Inoue, Manipulating γ-cyclodextrin-mediated photocyclodimerization of anthracenecarboxylate by wavelength, temperature, solvent and host, *Photochem. Photobiol. Sci.* **13**, 190–198 (2014); (b) Q. Wang, C. Yang, C. Ke, G. Fukuhara, T. Mori, Y. Liu, and Y. Inoue, Wavelength-controlled supramolecular photocyclodimerization of anthracenecarboxylate mediated by γ-cyclodextrins, *Chem. Commun.* **47**, 6849–6851 (2011).

71. C. Ke, C. Yang, T. Mori, T. Wada, Y. Liu, and Y. Inoue, Catalytic enantiodifferentiating photocyclodimerization of 2-anthracenecarboxylic acid mediated by a non-sensitizing chiral metallosupramolecular host, *Angew. Chem., Int. Ed.* **48**, 6675–6677 (2009).

72. (a) M. Rao, K. Kanagaraj, C. Fan, J. Ji, C. Xiao, X. Wei, W. Wu, and C. Yang, Photocatalytic supramolecular enantiodifferentiating dimerization of 2-anthracenecarboxylic acid through triplet–triplet annihilation, *Org. Lett.* **20**, 1680–1683 (2018); (b) W. Xu, W. Liang, W. Wu, C. Fan, M. Rao, D. Su, Z. Zhong, and C. Yang, Supramolecular assembly-improved triplet–triplet annihilation upconversion in aqueous solution, *Chem. A – Eur. J.* **24**, 16677–16685 (2018).

73. (a) S. Yi, V. Brega, B. Captain, and A. E. Kaifer, Sulfate-templated self-assembly of new M4L6 tetrahedral metal organic cages, *Chem. Commun.* **48**, 10295–10297 (2012); (b) R. J. Kuppler, D. J. Timmons, Q. R. Fang, J. R. Li, T. A. Makal, M. D. Young, D. Yuan, D. Zhao, W. Zhuang, and H. C. Zhou, Potential applications of metal-organic frameworks, *Coord. Chem. Rev.* **253**, 3042–3066 (2009).

74. (a) J. L. C. Rowsell and O. M. Yaghi, Metal–organic frameworks: A new class of porous materials, *Micropor. Mesopor. Mat.* **73**, 3–14 (2004); (b) A. R. Millward and O. M. Yaghi, Metal–organic frameworks with exceptionally high capacity for storage of carbon dioxide at room temperature, *J. Am. Chem. Soc.* **127**, 17998–17999 (2005).
75. M. Yoshizawa, Y. Takeyama, T. Kusukawa, and M. Fujita, Cavity-directed, highly stereoselective [2+2] photodimerization of olefins within self-assembled coordination cages, *Angew. Chem. Int. Ed.* **41**, 1347–1349 (2002).
76. Y. Nishioka, T. Yamaguchi, M. Kawano, and M. Fujita, Asymmetric [2 + 2] olefin cross photoaddition in a self-assembled host with remote chiral auxiliaries, *J. Am. Chem. Soc.* **130**, 8160–8161 (2008).
77. M. Yoshizawa, M. Tamura, and M. Fujita, Diels-alder in aqueous molecular hosts: Unusual regioselectivity and efficient catalysis, *Science* **312**, 251–254 (2006).
78. T. Kusukawa, M. Yoshizawa, and M. Fujita, Probing guest geometry and dynamics through host-guest interactions this work was supported by the CREST (core research for evolutional science and technology) project of Japan science and technology corporation, *Angew. Chem. Int. Ed.* **40**, 1879–1884 (2001).
79. M. Yoshizawa, Y. Takeyama, T. Okano, and M. Fujita, Cavity-directed synthesis within a self-assembled coordination cage: Highly selective [2 + 2] cross-photodimerization of olefins, *J. Am. Chem. Soc.* **125**, 3243–3247 (2003).
80. J. S. Seo, D. Whang, H. Lee, S. I. Jun, J. Oh, Y. J. Jeon, and K. Kim, A homochiral metal–organic porous material for enantioselective separation and catalysis, *Nature* **404**, 982–986 (2000).
81. M. Alagesan, K. Kanagaraj, S. Wan, H. Sun, D. Su, Z. Zhong, D. Zhou, W. Wu, G. Gao, H. Zhang, and C. Yang, Enantiodifferentiating [4 + 4] photocyclodimerization of 2-anthracenecarboxylate mediated by a self-assembled iron tetrahedral coordination cage, *J. Photochem. Photobiol. A: Chem.* **331**, 95–101 (2016).
82. R. Maeda, T. Wada, T. Mori, S. Kono, N. Kanomata, and Y. Inoue, Planar-to-planar chirality transfer in the excited state. Enantiodifferentiating photoisomerization of cyclooctenes sensitized by planar-chiral paracyclophane, *J. Am. Chem. Soc.* **133**, 10379–10381 (2011).
83. R. Joseph, A. Naugolny, M. Feldman, I. M. Herzog, M. Fridman, and Y. Cohen, Cationic pillararenes potently inhibit biofilm formation without affecting bacterial growth and viability, *J. Am. Chem. Soc.* **138**, 754–757 (2016).
84. G. Yu, J. Zhou, J. Shen, G. Tang, and F. Huang, Cationic pillar[6]arene/ATP host–guest recognition: Selectivity, inhibition of ATP hydrolysis, and application in multidrug resistance treatment, *Chem. Sci.* **7**, 4073–4078 (2016).

85. J.-C. Gui, Z.-Q. Yan, Y. Peng, J.-G. Yi, D.-Y. Zhou, D. Su, Z.-H. Zhong, G.-W. Gao, W.-H. Wu, and C. Yang, Enhanced head-to-head photodimers in the photocyclodimerization of anthracenecarboxylic acid with a cationic pillar[6]arene, *Chin. Chem. Lett.* **27**, 1017–1021 (2016).
86. S. Nakai, H. Sunayama, Y. Kitayama, M. Nishijima, T. Wada, Y. Inoue, and T. Takeuchi, Regioselective molecularly imprinted reaction field for [4 + 4] photocyclodimerization of 2-anthracenecarboxylic acid, *Langmuir* **33**, 2103–2108 (2017).
87. K. Bando, T. Zako, M. Sakono, M. Maeda, T. Wada, M. Nishijima, G. Fukuhara, C. Yang, T. Mori, T. C. S. Pace, C. Bohne, and Y. Inoue, Bio-supramolecular photochirogenesis with molecular chaperone: Enantio-differentiating photocyclodimerization of 2-anthracenecarboxylate mediated by prefoldin, *Photochem. Photobiol. Sci.* **9**, 655–660 (2010).
88. M. Nishijima, J.-W. Chang, C. Yang, G. Fukuhara, T. Mori, and Y. Inoue, Chiral recognition and supramolecular photoreaction of 1,1′-binaphthol with bovine and human serum albumins, *Res Chem. Intermed.* **39**, 371–383 (2013).
89. M. Nishijima, H. Tanaka, G. Fukuhara, C. Yang, T. Mori, V. Babenko, W. Dzwolak, and Y. Inoue, Supramolecular photochirogenesis with functional amyloid superstructures, *Chem. Commun.* **49**, 8916–8918 (2013).
90. M. Nishijima, H. Kato, G. Fukuhara, C. Yang, T. Mori, T. Maruyama, M. Otagiri, and Y. Inoue, Photochirogenesis with mutant human serum albumins: Enantiodifferentiating photocyclodimerization of 2-anthracenecarboxylate, *Chem. Commun.* **49**, 7433–7435 (2013).
91. M. Nishijima, M. Goto, M. Fujikawa, C. Yang, T. Mori, T. Wada, and Y. Inoue, Mammalian serum albumins as a chiral mediator library for bio-supramolecular photochirogenesis: Optimizing enantiodifferentiating photocyclodimerization of 2-anthracenecarboxylate, *Chem. Commun.* **50**, 14082–14085 (2014).
92. A. Joy, R. J. Robbins, K. Pitchumani, and V. Ramamurthy, Asymmetrically modified zeolite as a medium for enantioselective photoreactions: Reactions from spin forbidden excited states, *Tetrahedron Lett.* **38**, 8825–8828 (1997).
93. A. Joy, S. Uppili, M. R. Netherton, J. R. Scheffer, and V. Ramamurthy, Designed catalyst for enantioselective Diels-Alder addition from a C2-symmetric chiral bis(oxazoline)-iron(III) complex, *J. Am. Chem. Soc.* **122**, 728–729 (2000).
94. K. C. W. Chong, J. Sivaguru, T. Shichi, Y. Yoshimi, V. Ramamurthy, and J. R. Scheffer, Use of chirally modified zeolites and crystals in photochemical asymmetric synthesis, *J. Am. Chem. Soc.* **124**, 2858–2859 (2002).

Chapter 4

Chiral Memory and Its Applications

Alessandro D'Urso[*] and Roberto Purrello[†]

Department of Chemical Sciences, University of Catania
Viale A. Doria 6, Catania 95125, Italy

[*]*adurso@unict.it*
[†]*rpurrello@unict.it*

Transfer of chirality to self-assembled species is an interesting phenomenon that allows the induction of additional functionality to the supramolecular architectures. Exploiting the role of chiral templates, we elucidated how to design supramolecular systems able to memorize chiral information. Moreover, using the right combination of thermodynamic and kinetic pathways, on–off switches of stored chirality can be obtained.

1. Introduction

Self-assembly of molecules, driven by non-covalent intermolecular interactions, is a versatile and convenient route for the development of new materials with desired properties for possible technological applications.[1–5] Interestingly, self-assembly mediated by templates allows for achieving aggregates with additional characteristics such as controlled size and shape, defined stoichiometry, specific sequence, and dimensionality.[6–9] Among the properties that can be imposed on the resulting aggregates, chirality is one of the most intriguing features that can be induced by using chiral templates.[10–14] In the previous chapters, we discussed the importance of chirality and stated that it is one of the most intriguing

149

properties of matter; we also speculated on the possibility of transferring chiral information to self-assembled species of achiral molecules. We pointed out that induction and regulation of the supramolecular chirality to the self-assembly of achiral molecules has been usually achieved by chemical and/or physical "perturbations". Even the choice of the chiral template plays a key role regarding structure, size, and functional groups, in order to govern the interaction between constituents of multi-component systems, exploiting different driving forces (electrostatic interactions, coordination ligands, hydrophobic effects) to build chiral non-covalent architectures in aqueous solution.[15-20] Many templates have been used by scientists to achieve the transfer of chirality in supramolecular systems such as polymeric or single molecules, or organic or inorganic compounds. However, as yet, in several example of a chiral system obtained using a chiral template, supramolecular chirality is lost after the removal of the chiral mold (Figure 1, routes A, B and C). This is a major limitation for many applications of these assemblies because their chirality is expressed only in the presence of the template, which could interfere with the function for which the species has been designed.

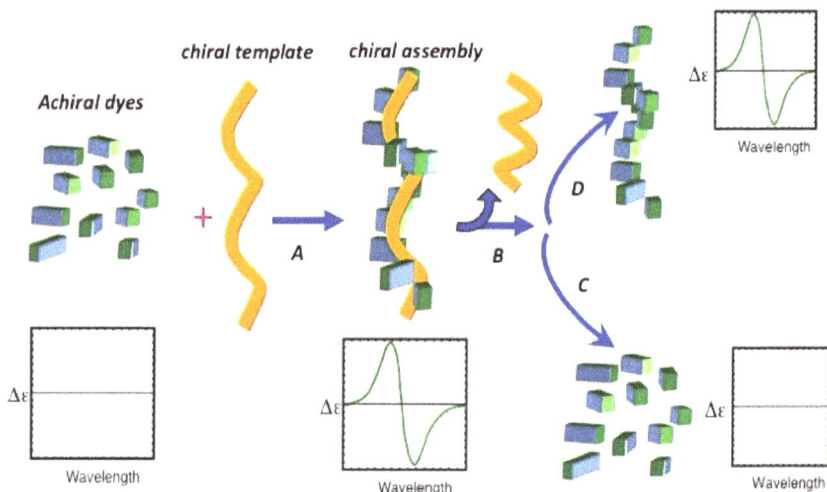

Figure 1.   Interaction of achiral dyes (no CD signal) with chiral template induces a CD signal in the absorption region of the dye (route A). Removal of the template (route B) causes the disappearance of the CD if the system is under thermodynamic control (route C). If the system is under kinetic control, then the assembly of the achiral dye maintains the chiral conformation (chiral memory, D).

In this chapter, we explore the possibility of obtaining supramolecular species in water, showing remarkable stability and inertness, allowing amplification and *memorization* of the chiral information imprinted by the chiral template. The success of this approach is based on the comprehension and exploitation of the hierarchical rules governing the interaction between constituents of multi-component systems. Even if some examples of memorization of chirality are present in the literature,[21–25] only few of these are performed in water. We focused our attention on the work where the solvent used is only water and organized the following several paragraphs to detail the different chiral templates exploited to gain the memorization of chirality.

2. Polymeric Templates

Polypeptides are simple scaffolds to design functional supramolecular systems as they have a number of favorable characteristics such as (1) large number of binding sites, (2) well-known and characteristic structures, (3) the possibility to tune the adopted conformations depending on the experimental conditions, and (4) large selection of functional groups that allow for the exploitation of different driving forces to induce interactions with molecules.

Among the building blocks for the formulation of functional supramolecular materials, porphyrins represent ideal compounds. They are stable under a variety of conditions, exhibit unique optical and redox properties, and can be readily functionalized at different sites. Charged substituents make porphyrins water soluble, even maintaining the hydrophobic character of the macrocyclic core. Exploiting this dichotomy, it is possible to drive the tendency of the porphyrins to self-assemble in aqueous solution and to modulate the aggregation onto other templates, inducing predefined architectures.[15–20] In this regard, the aggregation of water-soluble porphyrins onto oppositely charged polymeric chiral templates represents not only quite a simple strategy that allows for building tailored supramolecular species but is also a direct way to guide chirality of porphyrins at a supramolecular level.

We demonstrated that, inducing the formation of hetero-aggregates with oppositely charged porphyrins in the presence of chiral template (poly-glutamate) in aqueous solution, it is possible to obtain supramolecular species showing remarkable stability and inertness, allowing amplification and *memorization* of the chiral information imprinted by the

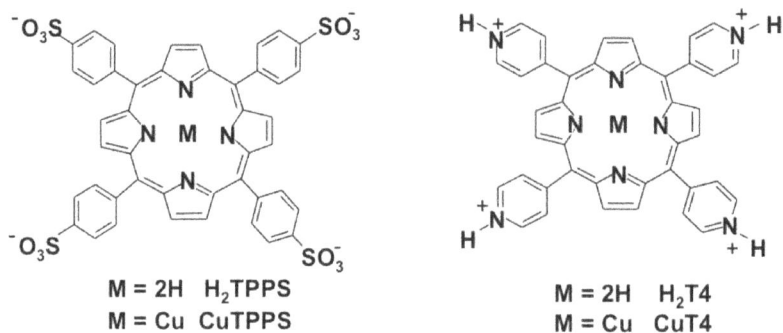

M = 2H   H₂TPPS
M = Cu   CuTPPS

M = 2H   H₂T4
M = Cu   CuT4

Figure 2. Schematic structures of the porphyrins tetracationic meso-tetrakis(N-methylpyridinium-4-yl)porphyrin (**H₂T4**) and copper(II) derivative (**CuT4**) and tetra-anionic meso-tetrakis(4-sulfonatophenyl)porphyrin (**H₂TPPS**) and copper(II) derivative (**CuTPPS**).

chiral template. In particular inducing the formation of a 1:1 hetero-aggregate between tetracationic meso-tetrakis(N-methylpyridinium-4-yl) porphyrin copper(II) (**CuT4**) and the tetra-anionic meso-tetrakis(4-sulfo-natophenyl)porphyrin (**H₂TPPS**) (Figure 2) in the presence of poly-L- or poly-D-glutamate, the hetero-aggregate gains chirality from the matrix.

This ternary system is able to retain the chiral information imprinted form the chiral mold, giving rise to the chiral memory phenomenon, as will be shown in what follows.[26,27] This behavior is due to the formation of an extended network of electrostatic interactions (i.e., those between the oppositely charged peripheral groups and the "π–π" between the porphyrin rings) that confer remarkable kinetic inertia to these ternary specie.[28] In detail, when an equimolar amount of **H₂TPPS** (in the pH range 3.2–4.0) is added to the preformed **CuT4**-poly-L-glutamate binary system, it induces drastic changes in the absorption, fluorescence, and CD spectra, indicating the formation of *chiral ternary* complexes.[26,27] The Soret bands of both porphyrins exhibit quite strong hypochromic effects (~50%), and the fluorescence emission of **H₄TPPS** is quenched by ~50% (**CuT4** is not fluorescent). Also, both the shape (bisignate) and the unusually high intensity of the induced CD features (Figure 3) strongly indicate that both porphyrins are extensively aggregated onto the α-helical poly-glutamate.[29] When the same complex is prepared using poly-D-glutamate, mirror images of the CD signals are detected (Figure 3), indicating that the chirality of these assemblies follows the matrix chirality.

These supramolecular ternary complexes behave quite differently compared to the parent *binary* species (poly-glutamate-**CuT4**). Indeed,

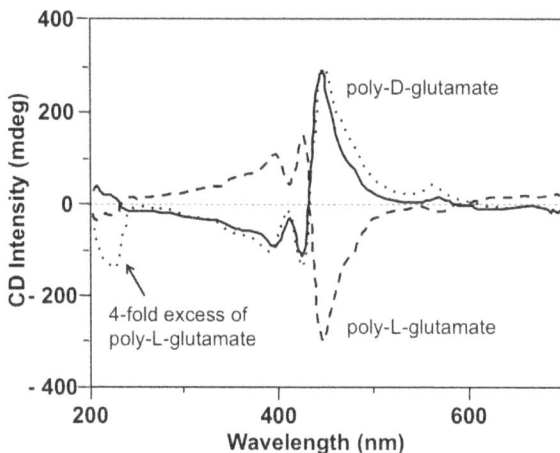

Figure 3. CD spectra of 4 μM solution of CuT4 in the presence of 200 μM poly-glutamate (solid curve with poly-L-glutamate and dashed curve with poly-D-glutamate) after the addition of H₂TPPS (4 μM). Dotted curve is the solution with poly-D-glutamate, 1 day after the addition of 4-fold excess poly-L-glutamate. Modified from Ref. [6].

although the **CuT4**-poly-glutamate binary system is chiral, it is kinetically *labile*. This is shown by the inversion (in about 10 minutes) of the induced CD band in the Soret region of **CuT4**-poly-D-glutamate upon the addition of a 4-fold excess amount of the L-form of the polymer. On the contrary, the addition of 4-fold excess of poly-L-glutamate to *ternary complexes* "built" on the poly-D-glutamate does *not* lead to the inversion of the induced CD signal in the Soret region (even after 5 *days* following the addition). The only indication of the L-form excess is the inversion of the helix marker bands at 222 and 208 nm (Figure 3). The same behavior has been observed for the ternary "L"-supramolecular species upon addition of an excess of poly-D-glutamate. The lack of inversion of the *induced* CD bands shows that, in contrast to their *binary* precursors, the poly-glutamate-**CuT4-H₄TPPS** aggregates are kinetically *inert*.

Analyzing the ICD behavior of the ternary complexes after inducing conformation transition of poly-glutamate by pH change, yet more proof of this notable inertia can be shown. Surprisingly, the ICD of the ternary complex persists after α-helix to random coil transition (from pH ~4 to 12) is induced. Indeed, one can think that since the chirality of the supramolecular complex is transferred from poly-glutamate to porphyrins, the pH-induced helix to coil transition should cause the disappearance of the ICD of the ternary species; however, increasing the pH to 12 does not

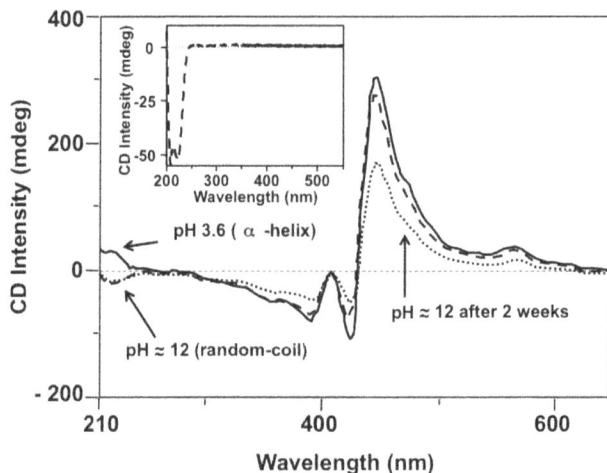

Figure 4.   CD spectra of CuT4 (4 μM) and H$_2$TPPS (4 μM) solution in the presence of poly-D-glutamate (200 μM) at pH 3.6 (solid curve) and at pH ~ 12 (dashed curve) and after 2 weeks (dotted curve). Inset: CD spectra of CuT4 (4 μM), H$_2$TPPS (4 μM) aggregated in water at pH 3.6 (solid curve) and after addition of poly-L-glutamate. Modified from Ref. [6].

perturb the ICD, strongly indicating that these porphyrin assemblies retain their "original" chirality even when the matrix loses it (Figure 4). Checking the time stability under such critical experimental conditions (pH 12), it turns out that these complexes remain stable for several days, as indicated by the CD intensity in the Soret region, which decreases only by 30% over about 4 weeks (Figure 4).

Notably, hetero-aggregation of the title porphyrins in ultra-pure water, in the absence of poly-glutamate, leads to achiral 1:1 complexes. This aggregation process is also accompanied by a hypochromic absorbance effect and fluorescence quenching similar to that observed in the presence of poly-glutamate, but no CD signal is observed (inset Figure 4). More interestingly, no dichroic signal is also observed in the Soret region if poly-glutamate is added after porphyrin self-aggregation (inset Figure 4). This behavior provides proof of the relevant kinetic inertia of these porphyrin supramolecular species and suggests that the role of the matrix is crucial only in the very first step of the formation of these ternary chiral species. In fact, once formed, these aggregates seem to have a "life" independent from the template but retain the "memory" of the shape of the "mold" used for their formation.

This is a good example of hierarchical control over the self-organization process and is related to the thermodynamic stability and kinetic inertia due to the interactions of convergent stabilization due to various kinds of electrostatic interactions.[30]

3. Aggregated Templates

In order to give an unambiguous demonstration of the memory phenomenon, the ICD signal in the Soret region should not be affected by the chiral mold removal. However, the removal experiment is not feasible when the mold is a covalent polymer because it could remain trapped in the porphyrin aggregates, which are quite extended species (about one million porphyrins for a single aggregate *molecule* as shown by elastic light scattering data).[26,27,31–33] To design an efficient removal experiment, the chiral template should exhibit several characteristics: (i) it should be a fluorescent monomeric chiral molecule, (ii) which, above a given concentration threshold, self-assembles in a chiral fashion, and (iii) establishing the equilibrium: n monomers $\leftrightarrow$ (monomer)$_n$.

We selected as a chiral template the "monomeric" Phenylalanine (Phe).[23] Indeed, it has a good fluorescence yield in a region where porphyrins only slightly absorb, and this allows us to monitor the residual concentration of the template after its removal from the solution much more efficiently compared with both the absorption and CD techniques. Moreover, the hydrophobic characteristic of a large portion of the Phe ensures greater aggregation tendency at lower concentration compared with other amino acids, leading to efficient induction of chirality.

Once the chiral template is chosen, the other issue is how is it possible to remove it from the solution. It has been demonstrated by dynamic light scattering (DLS) of Phe solution (Figure 5) that at 30°C there are large Phe aggregates (~60 nm) whose size decreases with increasing temperature, leveling off over 60°C (~20 nm).[23] This information allow us to perform ultrafiltration to remove the Phe aggregates from the solution. Even immediately following removal of the monomeric form of the "template" (e.g., by ultrafiltration), the equilibrium n monomers $\leftrightarrow$ (monomer)$_n$ shifts toward the left and the aggregated form disappears even if traces of the monomers (detail in what follows the threshold) remain in solution.

Then, we analyzed if Phe is able to induce chirality to porphyrin hetero-aggregates. Addition of the cationic **CuT4** and anionic **H₂TPPS** to an aqueous solution of L-Phe or D-Phe leads to an ICD in the Soret region

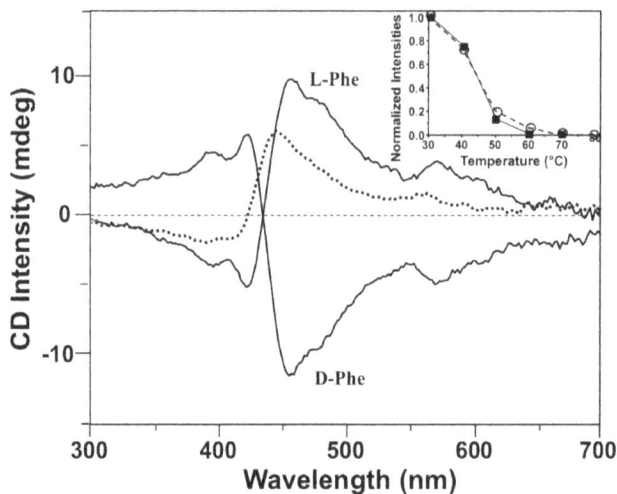

Figure 5.   CD spectra of H2TPPS and CuT4 (2×10^{-6} M each) in ultrapure Millipore water in the presence of L- or D-phenylalanine 8×10^{-3} M (solid curves), and after the removal of the L-enantiomer. (dotted curve). Inset shows intensity variations with temperature of ELS of 80 mM L-Phe solutions (circles) and CD in the Soret region of the imprinted CuT4-H2TPPS assemblies (squares). Modified from Ref. [11].

(Figure 5). The threshold concentration of Phe to induce the ICD is about 1 mM. Therefore, in contrast to all the previous examples, in this case the preferential conformation of the porphyrin aggregates is borrowed by chiral *non-covalent* polymers.

By increasing the temperature, the ICD decreases, paralleling the reduction of the mass average molecular weight, Mw, of Phe (Figure 5). At temperatures above 60°C, the ICD is reduced almost to the level of noise, suggesting that there are critical Phe cluster sizes (about 20 nm) and concentrations needed to initiate chiral porphyrin aggregation. In fact, porphyrin assembly formation is not affected by high temperature. Absorption and resonance light scattering (RLS) measurements confirm that at 80°C **CuT4-H$_2$TPPS** aggregates still form.[23]

According to the previous results,[26,27,31–33] the title porphyrin aggregates are inert enough to memorize the chirality of polymeric helical templates even after helix disruption. A similar result is expected here as well upon Phe removal. Ultrafiltration of the solution leaves, in fact, the CD signal almost unaltered (Figure 5), showing that the imprinted aggregates are now intrinsically chiral (they are so stable and inert that their memory lasts for many years).[23,31–33]

By fluorescence analysis of the solution, we assessed the residual concentration of Phe. Fluorescence has remarkable sensitivity and detects concentrations of samples as low as 1×10^{-8} M. DLS measurements have shown that this concentration is far below the concentration threshold necessary to transfer chirality to the porphyrin aggregates (about 10^{-3} M).[23]

A direct, fascinating consequence of the "memory" phenomenon is that the title aggregates are inherently chiral and thus, in principle, excellent templates for their self-replication. Indeed, when equimolar amounts of **CuT4** and **H₂TPPS** are individually added to solutions containing about 6×10^{-13} M of imprinted assemblies (each of them formed by about 2×10^6 porphyrin molecules, see Ref. [11]), the ICD of the imprinted aggregates increases linearly with porphyrin concentration, indicating that the chiral growth process is almost 100% enantiospecific.

These results suggest that the chiral self-growth rate should be much faster than porphyrin assembly as achiral species when they are further away from the chiral template. Indeed, kinetic investigations[24] have permitted the understanding of the discrimination mechanism and allowed to hypothesize the possible sequence of events hereafter reported: (i) after the initial (within the first 10 ms) random distribution of the porphyrin along the template surface, a second process (within 0.2 s) may lead to a conformational change of the template, which then favors the formation of a more discrete distribution of porphyrins on the template. It is only after this event that the interaction of a second porphyrin at topologically distinct sites may occur. At this point, the interaction of the second porphyrin follows the same sequence of events, followed by a slow hetero-aggregation of the two porphyrins, driven by electrostatic interactions and characterized by a strong chirality, as suggested by CD. These kinetic studies show that sub-picomolar concentrations of chiral non-covalent polymers play an efficient catalytic role in the aggregation of oppositely charged porphyrins. The rate of hetero-aggregation onto the chiral surface is, in fact, about two orders of magnitude faster than that measured in its absence, explaining the chiral amplification exerted by porphyrin aggregates. Considering the whole process, the rate-limiting step is the diffusion of the two self-aggregates of **CuT4** and **H₂TPPS** (bound at topologically distinct sites) on the Phe cluster surface to form chiral hetero-aggregates. When the chiral hetero-aggregates are used as a mold, the growth kinetics of self-similar species, as a result of the consecutive addition of the two porphyrins, is too fast to be followed using stopped-flow technique (data not published). This happens because the above-mentioned "slow" diffusion step on the template surface is no longer verifiable.

3.1. *On–Off switch: The memorized chirality*

As mentioned in previous paragraphs, the extended network of (strong and weak) electrostatic interactions between the tetracationic and tetra-anionic porphyrins lead to kinetic inertness and stability of these hetero-aggregates. Although this provides a great resource to store the chiral information, it represents a static system in opposition to the fundamentals of supramolecular chemistry, which is, on the contrary, based on dynamic and reversible behavior.

Therefore, we have designed a dynamic system by selecting porphyrins bearing the net charges in the peripheral substituents as being switchable on and off, allowing to modulate the network of electrostatic interactions that hold up the non-covalent supramolecular structure.[24]

To achieve this goal, we chose porphyrin meso-tetrakis(pyridinium-4-yl)porphyrin (**H₂TPyP**) with four piridyl groups that are ionizable (Figure 6 $pK_a \approx 4$) in the *meso* positions and can cycle between neutral (deprotonated) and cationic (protonated) states instead of four N-methylpiridyl groups (permanently cationic). In order to avoid nitrogen protonation of the inner core of **H₂TPPS** ($pK_a \approx 5$), we have used **CuTPPS** instead of the naked anionic parent (Figure 6). L-Phe and D-Phe are the chiral templates. Switching of the electrostatic interactions can be easily achieved by changing the pH; indeed, deprotonation of the four pyridines causes the loss of the positive charges (Figure 6) and, consequently, the disassembly of the supramolecular complex; on the contrary, re-protonation would lead to re-assembly of the aggregate.

Once the design of switchable hetero-aggregate has been overcome, the next issue is the preservation of chiral information. Indeed, to re-assemble the chiral supramolecular structure after chiral template removal is the crucial and most tricky step of the whole "erase and rewrite" process. Apparently, "switching on" the electrostatic interactions by re-protonating the four *meso* basic groups (Figure 6) should lead to the formation of an **achiral** complex between the two porphyrins because the chiral information should be lost after the erase reaction (Figure 6, route *a*).

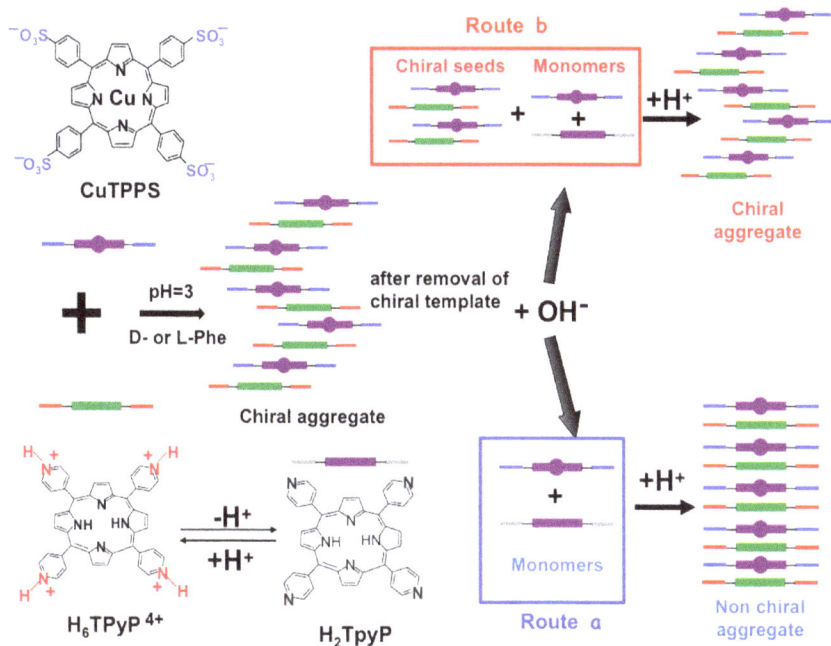

Figure 6. Schematic representation of porphyrin structures and of the possible two re-assembly pathways after disassembly induced by pH changes of the chiral porphyrin aggregate. Route *a*: The chiral aggregate is completely disassembled after the pH-jump. Route *b*: Some undetectable chiral seed survive the pH-jump.

However, the kinetic properties of these species suggest that it is possible to store/erase/rewrite the memory of chirality. As previously shown, the porphyrin hetero-aggregates are kinetically inert; this means that deprotonation of the four acid groups of the pyridil moieties would not cause the complete destruction of the chiral complexes but should leave a small amount of chiral seeds (Figure 6, route *b*). Since these species are huge (built by one million porphyrins), it is conceivable that most of the water is excluded from the inner part of these species, giving rise to families of pK_a values. As a result, when the pH is around the pK_a value we do not obtain complete disassembly since a certain small amount of aggregates might resist disassembly. In addition, these assemblies are excellent chiral templates that very efficiently self-catalyze their enantioselective growth.[24] Therefore, a long-lasting small amount of chiral seeds is enough to promote the re-formation of the **chiral** supramolecular complex (Figure 6, route *b*).

We define systems like this as **quasi-dynamic**. It is strictly related to the simultaneous presence of both chiral seeds in solution (static condition) and the monomers (dynamic condition). From a spectroscopic point of view, at high pH, the "ICD" signal disappears (in solution, there are achiral monomers plus a spectroscopically undetectable amount of chiral seeds) and reappears at low pH (chiral re-assembly) because the chiral re-assembly is driven by the seeds (amplification process).

As discussed in the previous paragraph, the non-covalent synthesis of the chiral porphyrin aggregate has been performed by adding equimolar amount of **H$_6$TPyP^{4+}** and **CuTPPS** to a solution (by HCl) of L- or D-Phe at pH 2.3, leading to mirror-images induced CD signal in the Soret region (inset A of Figure 7) and to about 90% of hypochromicity of the Soret band (inset B of Figure 7). It is worth noting that no CD signal is observed when the two porphyrins are mixed in pure water (pH 2.3). When the pH of the solution is increased from 2.3 to 9.0, the CD of the chiral porphyrin aggregates disappears (inset A of Figure 7) and the absorption spectra experience quite a strong hyperchromicity (inset B of Figure 7), indicating that most of the porphyrins are in a monomeric state. Decreasing the pH

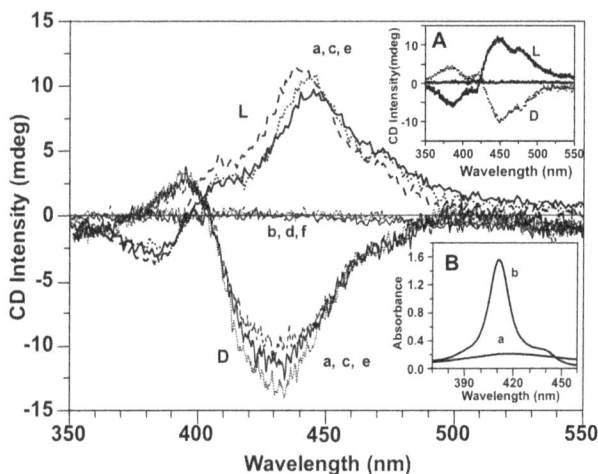

Figure 7.   CD spectra of porphyrin aggregate at pH 2.3 (a, c, e) and 9 (b, d, f) for the L- and D-imprinted aggregates. Inset A shows the CD spectra of aqueous solution of H2TpyP-CuTPPS (4 mM each) aggregated in the presence of L- or D-Phe. Inset B demonstrates the absorption spectra at pH 2.3 (a) and 9 (b), respectively. Modified from Ref. [13].

to 2.3 again, the CD signal of the porphyrin aggregate is restored (after about ten minutes). So, the system responds quite well to the external stimuli and, in the presence of phenylalanine, it is able to "indefinitely" switch off and on the supramolecular chiral information.

At this point, the next steps are (i) removal of the amino acid from the imprinted porphyrin hetero-aggregate solution (in acidic condition) and (ii) changing the pH from acid to basic and *vice versa* to "erase and rewrite" the CD signal (i.e., the chiral information).

Elimination of phenylalanine, by ultra-filtration of the solution, does not affect the chirality of the aggregates, as shown by the persistence of the CD signal in the Soret region (Figure 7 curves a). Also, further addition of the two porphyrins to the solution containing the supramolecular chiral assembly leads to the growth of the chiral architectures, confirming the remarkable ability of these chiral aggregates to template their own enantiospecific self-propagation.[36]

Then to "erase and rewrite" the chiral information of the amino acid-free chiral porphyrin assemblies, one has to change the pH conditions cyclically and monitor the ICD signal. The experiment shown in Figure 7 demonstrates that, indeed, the "ICD" of the phenylalanine-free chiral aggregate can be cyclically switched "on and off". The CD for both the L- and D-imprinted aggregates disappears by increasing the pH from 2.3 to 9.0 (Figure 7, curves b, d, f), and it is restored by lowering back the pH to 2.3 (Figure 7, curves a, c, e). No "ICD" is observed for the achiral aggregates when cycling between pH 2.3 and 9.0. Again, CD changes are parallel to the remarkable absorption variations. Protonation at pH 2.3 of the **H₂TPyP** peripheral nitrogens leads to aggregation and to about 90% of hypochromicity, whereas deprotonation (at pH 9.0) restores the Soret intensity (inset B of Figure 7). Ten consecutive cycles were performed to test the system on–off cycling ability, but, in principle, there are no apparent limitations to the number of cycles that can be performed.

The success of this experiment can be ascribed to the remarkable kinetic inertia of the imprinted aggregates that allows for the persistence of a spectroscopically undetectable amount of chiral seeds that drives the re-assembly of the chiral structure. Therefore, the re-assembly process should be time dependent. Indeed, if the chiral re-assembly is driven by the presence of inert chiral seeds, then (at pH 9.0) there must be a time interval after which the chiral seeds will disassemble. Then, chirality would not be reversible anymore and the system will re-assemble in a non-chiral fashion (see route *a* in Figure 6). Indeed, after about 24 hours at pH 9, the CD at pH 2.3 is not restored anymore because the chiral seeds also disassembled.

4. Inorganic Templates

In order to know how small the template can be to transfer the chiral information, monomeric "small" molecules showing chiral conformation were used as chiral templates, i.e, the Λ and Δ enantiomers of ruthenium (II) cationic complexes (≈ 9 Å, Figure 8).[35,36] This time, the chiral template is cationic and, in particular, the well-characterized cationic complex of ruthenium(II) with phenanthroline, $[Ru(phen)_3]^{2+}$ (Figure 8), which offers the possibility to activate the energy transfer processes from

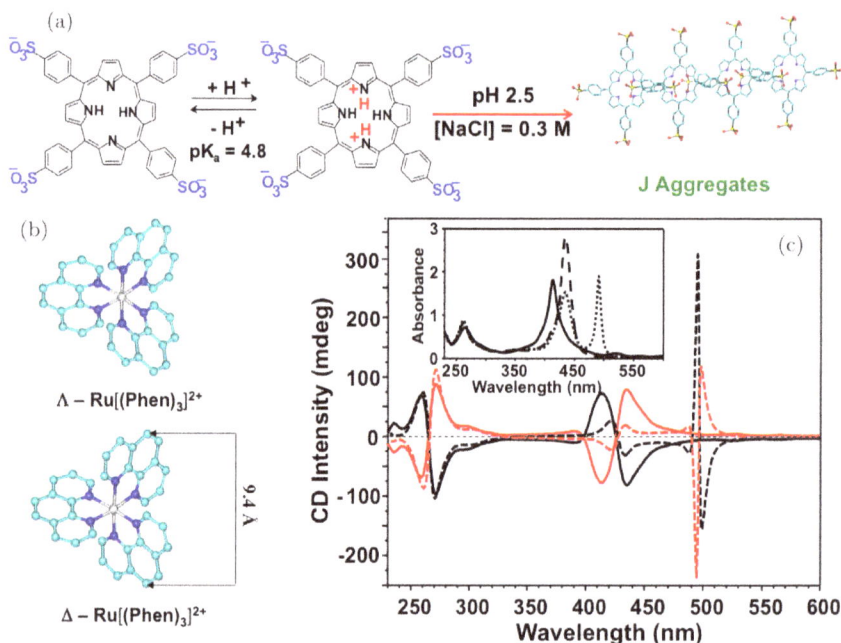

Figure 8.   Schematic representation of porphyrin structures and protonation-aggregation pathway (a) and $Ru[(Phen)_3]^{2+}$ enantiomers (b). (c) CD spectra of H_2TPPS ($10\,\mu M$) at pH 6.0, NaCl (0.3M) in the presence of Λ- (black curves) and Δ-$[Ru(phen)_3]^{2+}$ (red curves, 10 μM) and after changing the pH to 2.5 (dashed curves). Inset shows the absorption spectra of $H_2TPPS/[Ru(phen)_3]^{2+}$ complex species at pH 6 (solid curve) and at pH 2.5 (dashed curve) after one hour. Modified from Ref. [14].

the inorganic to the organic moiety.[37] As the organic part, in fact, was used tetra-anionic **H₂TPPS** (Figures 2 and 8), which has absorption bands (Q bands) in the region where the title ruthenium cationic complexes emit (around 600 nm).

As mentioned earlier, to memorize the chiral information the porphyrin aggregate must exhibit kinetic inertness; however, H₂TPPS at pH 6 does not aggregate in water solution; only at acidic pH (pKa of porphyrin core is ~ 4.8) does the **H₂TPPS** become zwitterionic and lead to extensive aggregation, forming J-aggregates (Figure 8). These aggregates are strongly stabilized by the network of both electrostatic and hydrophobic interactions and thus kinetically inert. The formation of J-aggregates requires particular experimental conditions (pH, ionic strength, concentration); therefore the preparation of the samples must follow a specific hierarchical order — a "wrong" sequence of instructions opens alternatives routes.

The first step of the synthesis is the interaction in aqueous solution at pH ≈ 6.0 (and in the presence of NaCl 0.3 M) of **H₂TPPS** and [Ru(phen)₃]²⁺. Interaction is evidenced by variation in the absorption spectra and, especially, by the appearance of an induced circular dichroism band in the absorption region of the achiral porphyrin (Figure 8). The relationship between chirality of the cationic metal complex and that transferred to the anionic porphyrins is straightforward because interactions of **H₂TPPS** with the Λ- and Δ-[Ru(phen)₃]²⁺ lead to mirror-image ICD signals (Figure 8).[14] The second and final step of the synthesis is accomplished by lowering the pH from ≈ 6.0 to 2.5. This leads to spectroscopic variations indicating the protonation of the central nitrogen atoms: that is, disappearance of the absorption band at 412 nm and appearance of the band of **H₄TPPS** at 436 nm (inset Figure 8). After around ten minutes, a new absorption band appears at 490 nm and shows the formation of the J-aggregates, (inset Figure 8). The CD spectrum changes accordingly: the exciton-coupled CD band centered at 420 nm disappears soon after the addition of HCl and, after around ten minutes, two new sets of exciton couplet bands centered at 422 nm (H aggregates) and 490 nm (J aggregates) appear (Figure 8). Also, in this case, it is evident that the chirality of the J-aggregates stems from the inorganic template because the exciton bands of H and J-aggregates formed with the two enantiomers are, in fact, mirror images of each other (Figure 8).[35,36]

In order to demonstrate the memorization of chirality since is not possible to remove the inorganic chiral template from the solution, we

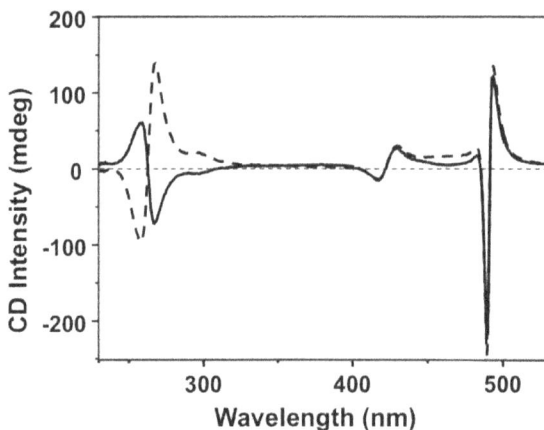

Figure 9.   CD spectra of the H₄TPPS J-aggregates formed in the presence of 10 μM Δ-[Ru(phen)₃]²⁺ before (dashed curve) and after (solid curve) the addition of an excess (15 μM) of Λ-[Ru(phen)3]2+.

performed addition of an excess of Δ-[Ru(phen)₃]²⁺ (15 μM) to J aggregates, in the presence of Λ-[Ru(phen)₃]²⁺ (10 μM). As a result, the addition of excess of the other [Ru(phen)₃]²⁺ enantiomers does not cause inversion of the CD bands in the visible region (where the absorption features of porphyrins dominate) but only in the UV region (the spectroscopic region where absorption of the inorganic complex occurs) (Figure 9). This straightforward experiment shows that this type of J-aggregates "remember" the chirality imprinted at the very onset of their formation.

In the next paragraph, we report that, using proper experimental conditions, the porphyrin J-aggregates in addition to the chiral memory phenomenon, offer even the possibility to cyclically erase and rewrite the chiral information, switching on and off the porphyrin ICD signal by pH variations.[35,36]

4.1. *Challenging the on–off switch*

In order to check the feasibility of switching Off and On the memory, we performed pH cycles (2.5 → 6.0 → 2.5) on a solution containing chiral J-aggregates of **H₄TPPS** templated on Λ-[Ru(phen)₃]²⁺ but in the presence of an excess of Δ-[Ru(phen)$_3$]$^{2+}$ (the system already discussed in Figure 9).

As shown in Figure 6, by changing the pH of the solution in these experimental conditions, two routes are possible (Figure 10). One (route A) leads to the enantiomeric form (Δ-[Ru(phen)$_3$]$^{2+}$/J-aggregates) of the starting aggregates (Λ-[Ru(phen)$_3$]$^{2+}$/J-aggregates), indicating that Λ-[Ru(phen)$_3$]$^{2+}$/J-aggregates are completely destroyed upon porphyrin deprotonation and thus the excess of the Δ enantiomer will manage the formation of J-aggregates with opposite chirality to that initially present. However, the second route (B) proceeds with retention of the starting chirality, demonstrating the remarkable inertness of the J-aggregates (having the ability of memorizing the chirality) which at pH 6 are only partially disassembled; therefore, a spectroscopically undetectable concentration of Λ-[Ru(phen)$_3$]$^{2+}$/J-aggregates remain in solution. Here again, the presence of a quasi-dynamic system is indispensable in order to obtain the reversible amplification of chiral undetectable seeds.

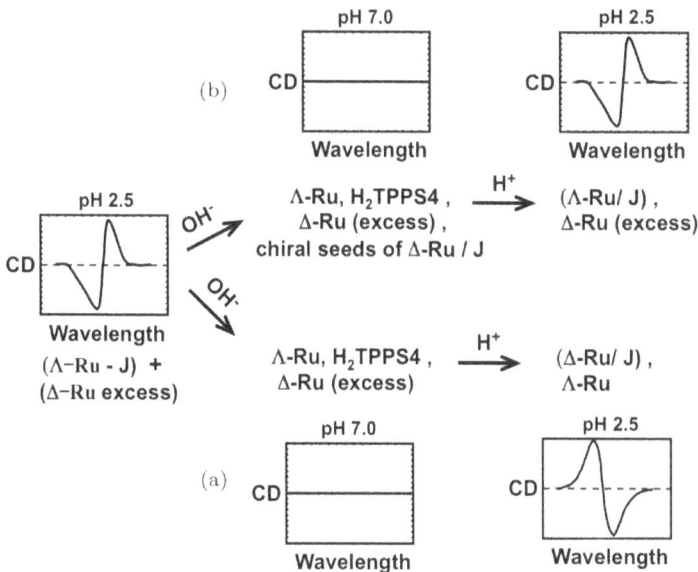

Figure 10. Scheme of the two possible re-assembly pathways following disassembly of the J-aggregates formed for interaction with Δ-[Ru-(phen)$_3$]$^{2+}$ but in the presence of an excess of Λ-[Ru(phen)$_3$]$^{2+}$. Modified from Ref. [14b].

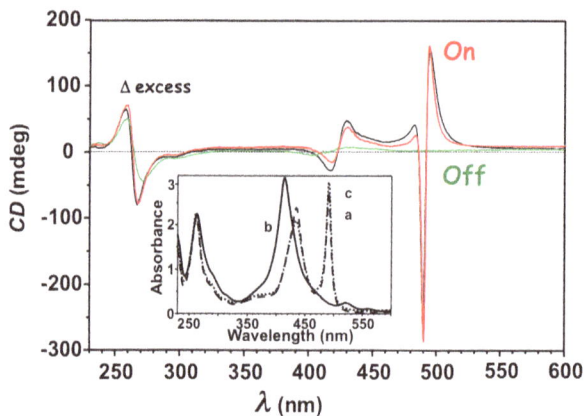

Figure 11.   CD spectra of the solution used for the measurements shown in Figure 9 recorded: (a) initially at pH 2.5 (black curve), (b) then at pH 6.0 (green curve), and (c) again at pH 2.5 (red curve). The inset shows the absorption spectra of the three experiments.

The experimental results show that B is the preferred route. Indeed, the CD spectra are consequently modified: when the pH is raised from 2.5 to around 6.0, the ICD of the J-aggregates disappears owing to deprotonation of H_2TPPS (Figure 11). As reported previously, in Figure 6, at pH 6 we should observe the CD signal of the (Δ-[Ru(phen)$_3$]$^{2+}$/H_2TPPS) complex; however, the formation of this complex is slow and the signal appears after 10–15 min. Then after the pH jump to 2.5, the CD signal of the J-aggregates is immediately restored (Figure 11). Remarkably, the sign of the exciton couplet is the same as that of the starting J complex, confirming the retention of chirality and the remarkable inertness of these chiral aggregates.[36] The same results have been obtained starting from the Δ-[Ru(phen)$_3$]$^{2+}$/J-aggregates.

The absorption spectra (inset Figure 11) show that, upon changing the pH value from 2.5 to around 6.0, the spectrum of the non-protonated H_2TPPS is restored: the subsequent decrease of pH to 2.5 leads again (within 5 min) to the J-aggregates.

Up to ten consecutive pH cycles were performed without observing any inversion of the CD Soret couplet. However, after about five pH cycles the solution becomes milky and scatters because of the formation of extended aggregates: this causes a decrease of the CD intensity. After these cycles, the same solution was kept at a pH of about 6.0 for 1, 2, 16, 24, and 48 h and 1 week. It should be noted that after each time interval when the pH value was lowered to 2.5, an identical CD signal shape was obtained, which shows the remarkable stability of these species.

These results can be explained in light of the experiment shown in the previous section: the inertness and catalytic properties of the initially formed aggregates (seeds), which allow the J species to retain the memory of chirality transferred from Λ- or Δ-[Ru(phen)$_3$]$^{2+}$. The increase of the pH value to around 6.0 does not disassemble all the aggregates: a residual concentration of these species remains in solution (the chiral seeds) and is able to very efficiently drive the correct folding of the memorized chirality despite the presence of an excess of the template that has an "opposite" handedness. On the contrary, the memory of chirality is lost if pH value is raised above 8. In these conditions, the chiral seeds are destroyed and porphyrin monomers redistribute on the conformers of the metal complex.

This example underlines the central role that hierarchy has in the non-covalent syntheses. It is, in fact, worth emphasizing that addition of chiral metal complexes to preformed J-aggregates or addition of all components (metal complex, porphyrins, acid, and salt) at pH 3 (we have tried all the different combinations) is not effective to obtain the system that is above-described.

5. Chiral Template that Spontaneously Dissociates

In the previous paragraphs, we showed that in order to demonstrate the storage of chiral information in self-assembled species, the chirality of the template has to be silenced. This goal can be obtained using several methods depending on the features of the template. For example, in the presence of polymeric templates it is possible to induce a conformational change of the matrix, which can lose or change the initial chirality,[26,27] or to add an excess of the template with opposite chirality.[35,36] In some cases, if the size of chiral template is restrained, the removal of the chiral mold can be achieved by time-consuming ultrafiltration procedure.[34] In this context, further studies on the possibility of spontaneous dissociation in aqueous solution of the chiral template, after induction of chirality, are highly desirable, in order to overcome the removal steps of the chiral template and to demonstrate the chiral memory phenomena.

These opportunities were offered by tetracoordinated Zn^{II} Schiff-base complexes, characterized by interesting photophysical and aggregation properties as well as by high synthetic versatility.[38–40] Actually, these complexes are Lewis acidic species capable of coordinating a large variety of donor Lewis bases;[38–44] therefore, it is expected that the coordinated water to the Lewis acidic Zn^{II} center enhances the Brønsted acidity,[45] thus generating hydroxide species, followed by demetallation.[46–51]

Indeed spectroscopic studies on aqueous solutions of ZnL_2-R and ZnL_2-S (Figure 12) through circular dichroism and UV–Vis absorption techniques indicate that these complexes are not stable in aqueous solution, undergoing almost complete dissociation after 24 hours. In detail, the UV–Vis spectra show a remarkable decrease of the absorption intensity and red shift of the two bands, at 295 nm and 365 nm, ascribable to Zinc complexes degradation (Figure 12). Also, circular dichroism spectra confirm that after 24 hours in aqueous solution the chirality of these zinc complexes disappears due to dissociation of the zinc complexes (Figure 12).

Initially, we demonstrated that it is possible to induce chirality to the porphyrin hetero-aggregates using the tetracationic **H₂T4** and the tetra-anionic **CuTPPS** derivative (Figure 2), mixing chiral templates such as chiral tetracoordinated Zn^{II} Schiff-base complexes, ZnL_2-R, and ZnL_2-S before their spontaneous dissociation,, in aqueous solution (Figure 13).[52] Porphyrin hetero-aggregates (**H₂T4–CuTPPS**) in aqueous solution can

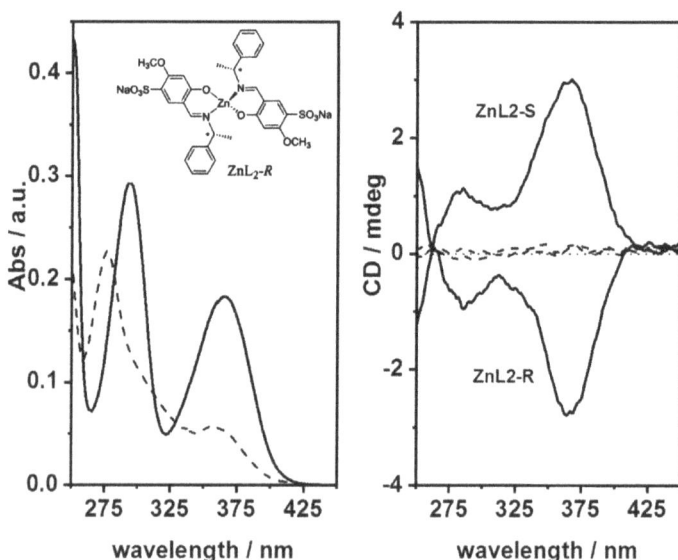

Figure 12.   UV and CD spectra of ZnL_2-R and ZnL_2-S complexes (10 μM) in water, recorded as dissolved in solution (solid curves) and after 24 h (dashed curves).

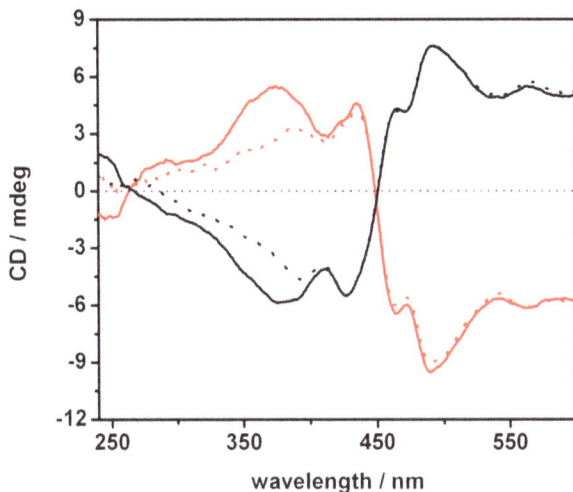

Figure 13. CD spectra of aqueous solution of porphyrin hetero-aggregates (H_2T4 6 μM and CuTPPS 6 μM) in the presence of ZnL_2 (10 μM) -R (black curves) and -S (red curves) enantiomers, as prepared (solid curves) and after 24h (dotted curves).

be prepared by mixing equimolar amounts of each porphyrins. In the absence of chiral templates i.e., ultra-clear conditions, porphyrin hetero-aggregates **H₂T4–CuTPPS** do not show supramolecular chirality. However, we achieved chiral porphyrin hetero-aggregates by adding equimolar amounts of each porphyrin to aqueous solutions of ZnL_2-*R* and ZnL_2-*S* complexes, leading to mirror-image ICD in the Soret region (Figure 13).

In order to demonstrate chiral memory phenomena, the solutions containing chiral porphyrin hetero-aggregates and chiral zinc complexes were stored for 24h, in order to accomplish the dissociation of the zinc complexes. According to previous results reported in literature on chiral memory phenomena, porphyrin hetero-aggregates are inert enough to memorize the chirality, and a direct consequence of this memory effect is that the aggregates keep the imprinted chirality by zinc complexes even after their dissociation. Indeed, the dissociation of chiral zinc complexes leads to disappearance of their CD at about 350 nm; on the contrary, the corresponding ICD signals of the imprinted hetero-aggregate do not change, indicating that porphyrin hetero-aggregate retains the memory of the template (Figure 13).

It is worth mentioning that mixing equimolar amounts of **H₂T4** and **CuTPPS** to a solution of ZnL_2-*S*, aged for 24 h, we found that porphyrin hetero-aggregates cannot show supramolecular chirality, underlined by zero ICD signal in Soret region (about 450 nm).

6. Conclusion

The results here reported confirm that memorization of chirality remains an intriguing and challenging phenomenon. Indeed, several points have to be focused on: (i) the chiral template should be easily removed from the solution, (ii) the building blocks selected have to form network of interactions (electrostatic, solvophobic, etc.) that trap aggregates at a quite deep local energy minimum, ensuring kinetic inertia, and (iii) there should be a possibility of switching between the imprinted and erased memory by changing the "nature" of the system component(s) through photo-, redox-, or pH-driven reactions without affecting the thermodynamic energy state of the chiral hetero-aggregates in order to get an erasable-rewritable (chiral) memory device.

References

1. J. M. Lehn, *Supramolecular Chemistry*, Wiley-VCH, Weinheim, (1995).
2. J. A. Elemans, A. E. Rowan, and R. J. Nolte, Mastering molecular matter. Supramolecular architectures by hierarchical self-assembly, *J. Mater. Chem.* **13**, 2661–2670 (2003).
3. O. Ikkala and G. ten Brinke, Functional materials based on self-assembly of polymeric supramolecules, *Science* **295**, 2407–2409 (2002).
4. D. Philip and J. F. Stoddart, Self-assembly in natural and unnatural systems, *Angew. Chem. Int. Ed. Engl.* **35**, 1154–1196 (1996).
5. T. Aida, E. W. Meijer, and S. I. Stupp, Functional supramolecular polymers, *Science* **335**, 813–817 (2012).
6. T. Yokoyama, S. Yokoyama, T. Kamikado, Y. Okuno, and S. Mashiko, Selective assembly on a surface of supramolecular aggregates with controlled size and shape, *Nature* **413**, 619–621 (2001).
7. B. Ferrer, G. Rogez, A. Credi, R. Ballardini, M. T. Gandolfi, V. Balzani, Y. Liu, H. R. Tseng, and J. F. Stoddart, Photoinduced electron flow in a self-assembling supramolecular extension cable, *Proc. Natl. Acad. Sci. U.S.A.* **103**, 18411–18416 (2006).
8. A. D'Urso, M. E. Fragalà, and R. Purrello, From self-assembly to noncovalent synthesis of programmable porphyrins' arrays in aqueous solution, *Chem. Commun.* **48**, 8165–8176 (2012).

9. T. E. Kaiser, H. Wang, V. Stepanenko, and F. Wurthner, Supramolecular construction of fluorescent J-aggregates based on hydrogen-bonded perylene dyes, *Angew. Chem. Int. Ed. Engl.* **46**(29), 5541–5544 (2007).

10. L. Zhang, T. Wang, Z. Shen, and M. Liu, Chiral nanoarchitectonics: Towards the design, self-assembly, and function of nanoscale chiral twists and helices, *Adv. Mater.* **28**, 1044–1059 (2016).

11. N. Micali, V. Villari, M. A. Castriciano, A. Romeo, and L. Monsù Scolaro, From fractal to nanorod porphyrin J-aggregates. Concentration-induced tuning of the aggregate size, *J. Phys. Chem. B.* **110**(16), 8289–8295 (2006).

12. Y. W. Chiang, R. M. Ho, E. L. Thomas, C. Burger, and B. S. Hsiao, A spring-like behavior of chiral block copolymer with helical nanostructure driven by crystallization, *Adv. Funct. Mater.* **19**(3), 448–459 (2009).

13. M. M. Smulders, A. P. Schenning, and E. Meijer, Insight into the mechanisms of cooperative self-assembly: The "Sergeants-and-Soldiers" principle of chiral and achiral C_3-symmetrical discotic triamides [*J. Am. Chem. Soc.* 2008, *130*, 606−611], *J. Am. Chem. Soc.*, **130**(12), 4204–4204 (2008).

14. M. Crego-Calama and D. N. Reinhoudt, *Supramolecular Chirality*, Springer, Berlin, (2006).

15. C. J. Medforth, Z. Wang, K. E. Martin, Y. Song, J. L. Jacobsen, and J. A. Shelnutt, Self-assembled porphyrin nanostructures, *Chem. Commun.* **47**, 7261–7277 (2009).

16. A. Romeo, M. A. Castriciano, I. Occhiuto, R. Zagami, R. F. Pasternack, and L. Monsù Scolaro, Kinetic control of chirality in porphyrin J-aggregates *J. Am. Chem. Soc.* **136**(1), 40–43 (2014).

17. T. van der Boom, R. T. Hayes, Y. Zhao, P. J. Bushard, E. A. Weiss, and M. R. Wasielewski, Charge transport in photofunctional nanoparticles self-assembled from zinc 5,10,15,20-tetrakis(perylenediimide)porphyrin building blocks, *J. Am. Chem. Soc.* **124**(32), 9582–9590 (2002).

18. H. L. Anderson, Building molecular wires from the colours of life: Conjugated porphyrin oligomers, *Chem. Commun.* **23**, 2323–2330 (1999).

19. G. A. Schick, I. C. Schreiman, R. W. Wagner, J. S. Lindsey and D. F. Bocian, Spectroscopic characterization of porphyrin monolayer assemblies, *J. Am. Chem. Soc.* **111**, 1344–1350 (1989).

20. A. Gulino, P. Mineo, E. Scamporrino, D. Vitalini, and I. Fragalà, Molecularly engineered silica surfaces with an assembled porphyrin monolayer as optical NO_2 molecular recognizers, *Chem. Mater.* **16**(10), 1838–1840 (2004).

21. F. Helmich, C. C. Lee, A. P. Schenning, and E. W. Meijer, Chiral memory via chiral amplification and selective depolymerization of porphyrin aggregates, *J. Am. Chem. Soc.* **132**(47), 16753–16755 (2010).

22. H. Onouchi, T. Miyagawa, K. Morino, and E. Yashima, Assisted formation of chiral porphyrin homoaggregates by an induced helical poly(phenylacetylene)

template and their chiral memory, *Angew. Chem. Int. Ed. Engl.* **45**(15), 2381–2384 (2006).

23. R. Lauceri, A. Raudino, L. M. Scolaro, N. Micali, and R. Purrello, From achiral porphyrins to template-imprinted chiral aggregates and further. Self-replication of chiral memory from scratch, *J. Am. Chem. Soc.* **124**(6), 894–895 (2002).

24. R. Lauceri, G. F. Fasciglione, A. D'Urso, S. Marini, R. Purrello, and M. Coletta, Kinetic investigation of porphyrin interaction with chiral templates reveals unexpected features of the induction and self-propagation mechanism of chiral memory, *J. Am. Chem. Soc.* **130**(32), 10476–10477 (2008).

25. K. Toyofuku, M. A. Alam, A. Tsuda, N. Fujita, S. Sakamoto, K. Yamaguchi, and T. Aida, Amplified chiral transformation through helical assembly, *Angew. Chem. Int. Ed. Engl.* **46**(34), 6476–6480 (2007).

26. E. Bellacchio, R. Lauceri, S. Gurrieri, L. Monsù Scolaro, A. Romeo, and R. Purrello, Template-imprinted chiral porphyrin aggregates, *J. Am. Chem. Soc.* **120**(47), 12353–12354 (1998).

27. R. Purrello, A. Raudino, L. Monsù Scolaro, A. Loisi, E. Bellacchio, and R. Lauceri, Ternary porphyrin aggregates and their chiral memory, *J. Phys. Chem. B* **104**(46), 10900–10908 (2000).

28. C. A. Hunter, J. K. M. Sanders, The nature of π–π interactions, *J. Am. Chem. Soc.* **112**, 5525–5534 (1990).

29. E. J. Gibbs, I. Tinoco, M. Maestre, P. A. Ellinas, and R. F. Pasternack, Self-assembly of porphyrins on nucleic acid templates. *Biochem. Biophys. Res. Commun.* **157**(1), 350–358 (1988).

30. R. Lauceria, M. De Napoli, A. Mammana, S. Nardis, A. Romeo, and R. Purrello, Hierarchical self-assembly of water-soluble porphyrins, *Synth. Met.* **147**(1–3), 49–55 (2004).

31. A. Mammana, M. De Napoli, R. Lauceri, and R. Purrello, Induction and memory of chirality in porphyrin hetero-aggregates: The role of the central metal ion, *Bioorg. Med. Chem.* **13**(17), 5159–5163 (2005).

32. R. Lauceri, R. Purrello, Transfer, memory and amplification of chirality in porphyrin aggregates, *Supramol. Chem.* **17**, 61–66 (2005).

33. R. Lauceri, A. D'Urso, A. Mammana, and R. Purrello, Chiral memory: Induction, amplification, and switching in porphyrin assemblies, *Chirality* **20**, 411–419 (2008).

34. A. Mammana, A. D'Urso, R. Lauceri, and R. Purrello, Switching off and on the supramolecular chiral memory in porphyrin assemblies, *J. Am. Chem. Soc.* **129**(26), 8062–8063 (2007).

35. R. Randazzo, R. Lauceri, A. Mammana, A. D'Urso, and R. Purrello, Interactions of Λ and Δ enantiomers of ruthenium(II) cationic complexes with achiral anionic porphyrins, *Chirality* **21**, 92–96 (2009).

36. R. Randazzo, A. Mammana, A. D'Urso, S. Lauceri, and R. Purrello, Reversible "chiral memory" in ruthenium tris(phenanthroline)–anionic porphyrin complexes, *Angew. Chem. Int. Ed.* **47**, 9879–9882 (2008).
37. V. Balzani and A. Juris, Photochemistry and photophysics of Ru(II)-polypyridine complexes in the Bologna group. From early studies to recent developments, *Coord. Chem. Rev.* **21**, 97–115 (2001).
38. G. Consiglio, S. Failla, P. Finocchiaro, I. P. Oliveri, and S. Di Bella, Aggregation properties of bis (salicylaldiminato) zinc (II) Schiff-base complexes and their Lewis acidic character, *Dalton Trans.* **41**(2), 387–395 (2012).
39. S. Di Bella, I. P Oliveri, A. Colombo, C. Dragonetti, S. Righetto, and D. Roberto, An unprecedented switching of the second-order nonlinear optical response in aggregate bis(salicylaldiminato)zinc(II) Schiff-base complexes, *Dalton Trans.* **41**(23), 7013–7016 (2012).
40. G. Consiglio, S. Failla, P. Finocchiaro, I. P. Oliveri, R. Purrello, and S. Di Bella, Supramolecular aggregation/deaggregation in amphiphilic dipolar Schiff-base zinc(II) complexes, *Inorg. Chem.* **49**(11), 5134–5142 (2010).
41. I. P. Oliveri, G. Malandrino, and S. Di Bella, Phase transition and vapochromism in molecular assemblies of a polymorphic zinc(II) Schiff-base complex, *Inorg. Chem.* **53**(18), 9771–9777 (2014).
42. I. P. Oliveri, G. Maccarrone, and S. Di Bella, A Lewis basicity scale in dichloromethane for amines and common nonprotogenic solvents using a zinc (II) Schiff-base complex as reference Lewis acid, *J. Org. Chem.* **76**(21), 8879–8884 (2011).
43. I. P. Oliveri and S. Di Bella, Sensitive fluorescent detection and Lewis basicity of aliphatic amines, *J. Phys. Chem. A* **115**(50), 14325–14330 (2011).
44. G. Consiglio, S. Failla, I. P. Oliveri, R. Purrello and S. Di Bella, Controlling the molecular aggregation. An amphiphilic Schiff-base zinc (II) complex as supramolecular fluorescent probe, *Dalton Trans.* **47**, 10426–10428 (2009).
45. G. Parkin, Synthetic analogues relevant to the structure and function of zinc enzymes, *Chem. Rev.* **104**(2), 699–767 (2004).
46. S. J. Wezenberg, E. C. Escudero-Adán, J. Benet-Buchholz, and A. W. Kleij, Colorimetric discrimination between important alkaloid nuclei mediated by a bis-salphen chromophore, *Org. Lett.* **10**(15), 3311–3314 (2008).
47. R. Novotná, Z. Trávníček, and I. Popa, Synthesis and characterization of the first zinc (II) complexes involving kinetin and its derivatives: X-ray structures of 2-chloro-N6-furfuryl-9-isopropyladenine and [Zn(kinetin)2Cl₂]·CH₃OH, *Inorg. Chim. Acta.* **363**(10), 2071–2079 (2010).
48. M. Enamullah, V. Vasylyeva, and C. Janiak, Chirality and diastereoselection of Δ/Λ-configured tetrahedral zinc (II) complexes with enantiopure or racemic Schiff base ligands, *Inorg. Chim. Acta.* **408**, 109–119 (2013).
49. A. C. Chamayou, S. Lüdeke, V. Brecht, T. B. Freedman, L. A. Nafie, and C. Janiak, Chirality and diastereoselection of Δ/Λ-configured tetrahedral zinc

complexes through enantiopure schiff base complexes: Combined vibrational circular dichroism, density functional theory, 1H NMR, and X-ray Structural Studies, *Inorg. Chem.* **50**(22), 11363–11374 (2011).

50. I. P. Oliveri, S. Failla, A. Colombo, C. Dragonetti, S. Righetto, and S. Di Bella, Synthesis, characterization, optical absorption/fluorescence spectroscopy, and second-order nonlinear optical properties of aggregate molecular architectures of unsymmetrical Schiff-base zinc (II) complexes, *Dalton Trans.* **43**(5), 2168–2175 (2014).

51. G. Consiglio, S. Failla, P. Finocchiaro, I. P. Oliveri, and S. Di Bella, An unprecedented structural interconversion in solution of aggregate zinc (II) salen Schiff-base complexes, *Inorg. Chem.* **51**(15), 8409–8418 (2012).

52. M. Gaeta, I. P. Oliveri, M. E. Fragalà, S. Failla, A. D'Urso, S. Di Bella, and R. Purrello, Chirality of self-assembled achiral porphyrins induced by chiral Zn(II) Schiff-base complexes and maintained after spontaneous dissociation of the template: A new case of chiral memory, *Chem. Commun.* **52**, 8518–8521 (2016).

Chapter 5

Role of Physical Bias on Chiral Induction: Hydrodynamic Flows and Temperature Gradients

Luigi Monsù Scolaro

Dipartimento di Scienze Chimiche, Biologiche, Farmaceutiche ed Ambientali, University of Messina, Viale F. Stagno d'Alcontres 31, Messina 98167, Italy

lmonsu@unime.it

Chirality in supramolecular chemistry is an intriguing property with potential applications. When considering achiral building components, chiral induction can be achieved through chemical or physical chiral bias. This chapter focuses on the effects of hydrodynamic and convective flows in expressing chirality at the supramolecular level, especially using porphyrin-based aggregated systems. The hydrodynamic flow in a vortex is able to sculpt the growing nanoaggregates leading to a strong statistical correlation between the rotation sense and the sign of the circular dichroism (CD) spectra. Vortexing allows also to separate enantiomorphous aggregates from a racemic mixture. Models of the hydrodynamic flow inside various types of vessels have been described in the literature. A strong magnetic field affords a way to control the effective gravity and, in combination with rotation, offers a clear demonstration that a falsely chiral bias (rotation and gravity) can lead to enantioselection in systems far from thermodynamic equilibrium. Recent results have shown that a proper design of microfluidic devices generates local microvortices able to imprint chirality into the growing aggregates. Finally, the effect of slight temperature gradients will be examined, revealing the chiroptical response in very flexible aggregates due to their orientation and deformation induced by convective flows inside the vessels.

1. Introduction

Chirality is a ubiquitous phenomenon in nature related to asymmetry or low symmetry and the difference existing between an object and its mirror image. Our hands represent a common example of an enantiomorphous pair, since, according to the static definition of Lord Kelvin, an object is chiral "if its image in a plane mirror, ideally realized, cannot be brought to coincide with itself".[1] The involvement of chirality at the molecular level and its further expression at higher levels is a current and largely investigated topic.[2,3] In biological systems, it plays an important role, and many relevant recognition mechanisms are based on the possibility to discriminate between molecules differing only in their handedness. In this context, homochirality is another very important issue for its implications in prebiotic chemistry and evolution of life itself. Life on earth is based on homochiral building blocks, i.e., L-aminoacids and D-sugars, to access all the complex biological systems, such as proteins, polysaccharides, and nucleic acids. Many hypotheses on the origin (terrestrial or extraterrestrial) of this phenomenon have been proposed as possible scenarios even if a commonly accepted mechanism is still lacking.[4–6] For example, CPL has been demonstrated as being effective in chiral resolution of racemic mixtures.[7,8] Indeed, CPL has been detected in star formation regions, even if in the near-IR region, that is too low in energy for effective chemical processes.[9] Vortex motion in galaxies or during the growth of low-mass stars has been also pointed as a possible font of chiral matter. Gas flow could generate opposite enantiomeric species with respect to the north and south regions of the rotating stellar disk.[10] Also, parity violation and consequently symmetry breaking has been demonstrated in physics,[11,12] but the energy difference calculated for two enantiomers is so small in comparison to the energy in chemical reactions (in the order of $10^{-10} J$ or even less) that its involvement in homochirality has been largely debated by the chemical community.[4]

Moving from the molecular to the supramolecular level, the expression of chirality has been intensely investigated: chiral transfer,

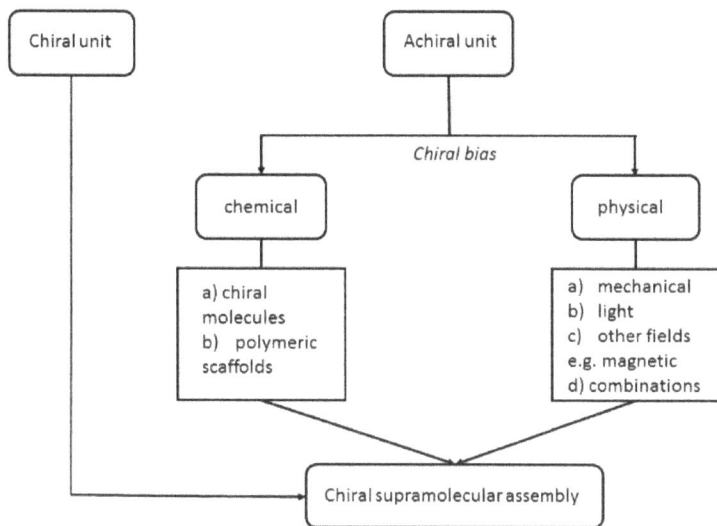

Figure 1.   General scheme to achieve a chiral supramolecular assembly starting from chiral or achiral building blocks.

amplification, and memory[13–19] have attracted interest since they have a close relationship with biological systems. The formation of supramolecular self-assemblies can be quite a complex thermodynamically or kinetically driven process,[20] and the chirality can be introduced at different points along the aggregation pathways. In any case, in order to achieve a selected chiral state, a chiral bias is needed, either at molecular or at higher level (Figure 1).

Accordingly, on one hand, when the basic building blocks are chiral, the appearance of chirality in the final supramolecular architecture is expected and commonly observed,[3] even if some level of complexity can emerge.[21] On the other hand, when the basic constituent units are achiral, the chirality can be transferred to the final system through chemical or physical biases.

In this respect, the role of achiral building blocks, which can self-assemble onto homochiral templates, either simple chiral molecules or more complex polymeric scaffolds, has received a lot of attention.[22–27] An open question is related to the unavoidable presence of the chiral templating agent, which could preclude a clear demonstration of the propagation and amplification steps. Anyway, much more intriguing is the case of spontaneous symmetry breaking in racemic mixtures induced by not-chemical chiral perturbations.

In order to better introduce this point, we have to make a distinction between true and false chirality. This idea has been proposed by L. Barron to implement the static concept of chirality attributed to Lord Kelvin, thus including dynamical effects. According to its original definition, "true chirality is exhibited by systems that exist in two distinct enantiomeric states that are interconverted by space inversion (parity $\hat{P}$), but not by motion reversal (time reversal $\hat{T}$) combined with any proper spatial rotation".[28] As a consequence, CPL or a magnetic field aligned parallel to unpolarized light is a truly chiral field (Figure 2), since space inversion leads to an enantiomorphous arrangement that cannot be obtained by the action of time reversal followed by rotation.

Accordingly, using synchrotron CPL, UV photolysis experiments on racemic mixtures of the amino acid leucine demonstrated effectiveness in obtaining enantioselection, affording 2.6% enantiomeric excess.[8] Even better, a combination of CPL and grinding has also been recently proposed as a successful method to obtain chiral selection.[29] Magnetochiral effect has been theoretically proposed by Wagnière[30] and later experimentally

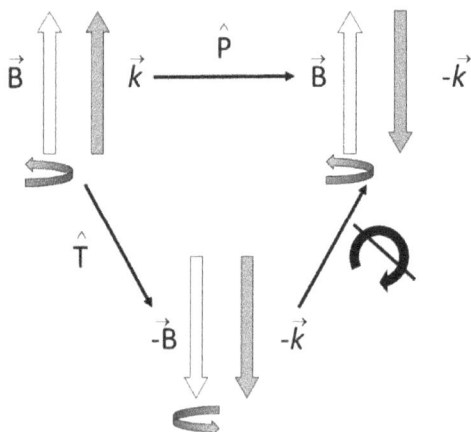

Figure 2.　The parallel combination of the magnetic field $\vec{B}$ and the light beam $\vec{k}$ ($\vec{B}$, $\vec{k}$) is converted by space inversion ($\hat{P}$) into an antiparallel orientation of the vectors ($\vec{B}$, $-\vec{k}$) that is not superimposable with the starting arrangement; time reversal ($\hat{T}$) followed by a 180° rotation on the initial combination does not afford the same final state, confirming that it is truly chiral, according to Barron's definition.

proved by Rikken and Raupach.[31,32] More recently, several theoretical and experimental investigations have pointed out that a hydrodynamic chiral perturbation is able either to induce chirality in a system[33] or to separate a racemic mixture.[34,35] In supramolecular chemistry, a growing number of examples, mainly based on self-assembled porphyrin structures, are available, demonstrating that chiral hydrodynamic flows generated by stirring,[36,37] or vortexing[38] the solutions during aggregation are able to lead to enantioselection. On the contrary, combination of parallel electric and magnetic fields or gravity and rotation are falsely chiral biases (Figure 3). In particular, the coupling of gravity and rotation has been highly debated in chemical reactions leading to equilibrium mixtures,[39] since a falsely chiral bias is theoretically expected to yield no effect in such conditions.[40]

Quite recently, the combination of rotational and magnetic forces (actually controlling the effective gravity acting on the samples) has provided a good evidence that in a system far from equilibrium and under kinetic control, even a falsely chiral bias can induce enantioselection.[41] The application of an external falsely chiral bias acts differently on the

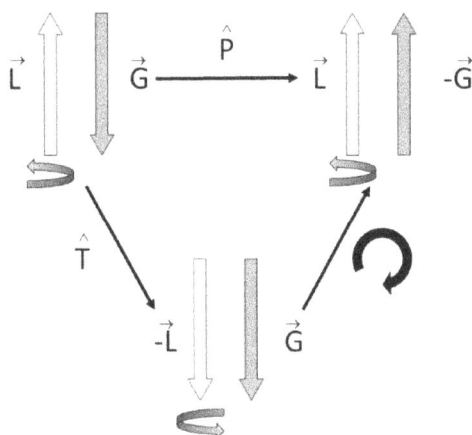

Figure 3.   The antiparallel combination of the angular momentum $\vec{L}$ and the gravity $\vec{G}$ ($\vec{L}$, $\vec{G}$) is converted by space inversion ($\hat{P}$) into a parallel orientation of the vectors ($\vec{L}$, $-\vec{G}$) that is not superimposable with the starting arrangement; anyway, the same arrangement is afforded by time reversal ($\hat{T}$) followed by a 180° rotation, confirming that the combination of these two vectors is falsely chiral, according to Barron's definition.

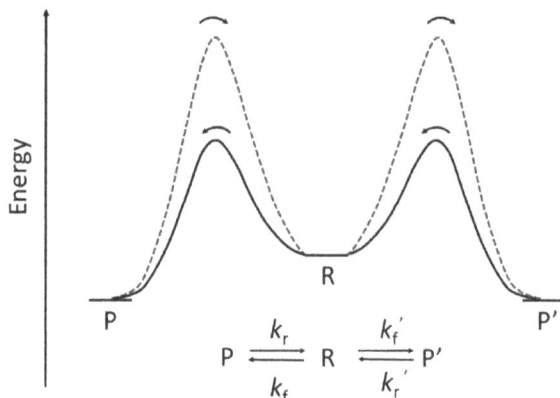

Figure 4.   Under a falsely chiral bias, the energy profiles for an achiral reactant R leading to enantiomeric product species P and P′ are altered, and the rates for the forward and the reverse pathway to the same enantiomer are different.

activation energies for the forward and the reverse reactions, providing a violation of the microscopic reversibility principle (Figure 4).

Even if apparently related to the hydrodynamics of the fluid, a completely different scenario occurs during crystallization in supersaturated solutions. Kondepudi and coll. showed that the spontaneous resolution of $NaClO_3$ crystals into optically pure species is a consequence of the stirring imposed to the solutions during the crystallization process with a magnetic bar. However, there is no statistical correlation between the observed handedness of the crystals and the rotation sense. The mechanical breaking of the initial crystallite into a large number of smaller nuclei, having a well-defined handedness, leads to a further growth of homochiral crystals.[42,43]

Another quite intriguing source of a physical bias is the thermophoretic force due to thermal gradients in the reaction vessels that can be chiral if there is a proper distribution of temperature values among the various walls. Even if difficult to control, this effect could lead to appearance of chirality in supramolecular assemblies, and some examples have been described in literature.[44–46]

This chapter is not intended to be exhaustive, since already a number of specialized review articles have been specifically dedicated to the subject.[47–49] Rather, some examples will be reported from the chemistry of porphyrins and related pigments on the role of vortexes and thermally induced hydrodynamic flows.

2. Hydrodynamic Flows

2.1. *General considerations*

Since the earlier and debated observations of optical activity in supramolecular assemblies of chromophores obtained under different stirring conditions, a growing number of reports have treated the topic of chirality as a consequence of hydrodynamic flows. The phenomena described till now have been referred to as mechanically induced and can be irreversible or reversible, depending on if they persist or not upon removal of the perturbation.

The size and shape, together with the elasticity of the supramolecular structure can determine the extent of the effect and its stability over time. A globular aggregate is less prone to deformation when compared to a ribbon-like structure. In this latter case, the nature of the applied flow can address the torsion toward a spiral or a helix (Figure 5). The energy request to force a ribbon into a helix (twisted ribbon) is higher with respect to that necessary to obtain a spiral (helicoidal ribbon). This difference could explain why the former structural motif leads to reversible effects (chirality disappears after removing the perturbation), while the latter is responsible for irreversible ones. Also the geometry of the vessel and the methodology for generating the hydrodynamic flow (rotation, mechanical, or magnetic stirring) are important parameters, since the chiral perturbation can change in intensity and invert its sign as a function of the position inside the solution (Figure 6). Another consequence of the

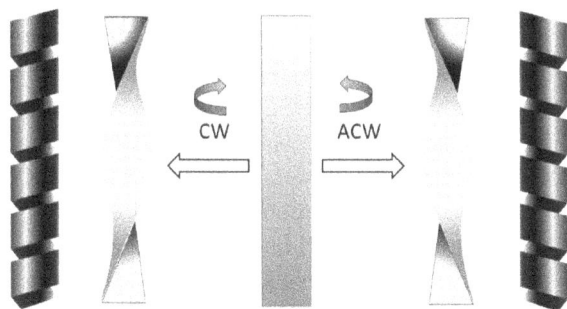

Figure 5.   The torque applied by a hydrodynamic flow can twist a nano-ribbon into a helix (or twisted ribbon) or into a spiral (or helicoidal ribbon). Clockwise (anticlockwise) rotation gives a right (left)-handed helix.

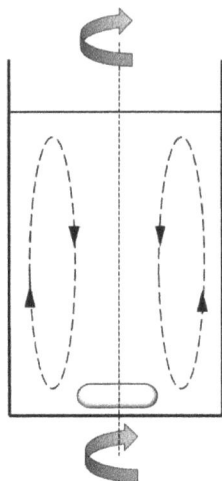

Figure 6.   In the case of a solution stirred by a magnetic bar inside a cell, the fluid rotates mainly along the perpendicular axis. The rotational flow rates are fast at the bottom and slower close to the top of the axes, with an ascending flow at the wall and a descending one in the center of the cell.

flow is the potential alignment of the particles, whose extent is again dependent on their shape and size. All these aspects have been widely discussed in several review articles by Ribò et coll.[47,48]

The detection of optical activity is traditionally performed by chemists measuring CD spectra using conventional spectropolarimeters. This approach caused most of the controversial results at the beginning of the interest toward this topic. The permanent or temporary deformation and alignment on the supramolecular aggregates can lead to true CD, which is a feature of chirality, or artifactual CD, arising from linear dichroism in aligned samples and generating misinterpretation of the experimental results. This problem has been solved by applying Mueller Matrix spectroscopy that allows to dissect the contribution of true CD, true circular birefringence, linear dichroism, and linear birefringence.[50]

A further problem in studying such systems is the importance of the mixing protocols in preparing the supramolecular assemblies. In the case of kinetically trapped (or metastable) non-dissipative non-equilibrium systems, a specific final aggregated state is reached only by adopting a precise protocol for mixing the various components, together with a fine control of the medium properties in terms of pH, ionic strength, and temperature.[20] The need for an accurate description of the experimental conditions and procedures has been recognized quite recently, explaining, in part, the difficulty in reproducing early results in different laboratories.

At this point, some fundamental questions need for answers. *How is chirality transferred from the physical perturbation to the supramolecular system? Is the flow acting at the beginning or along the aggregation process?* The two questions are somehow connected and still they have a strong impact on the search for homochirality. Different scenarios are possible depending on the nature of the supramolecular aggregation process, i.e., if it is at equilibrium or not. In a system at thermodynamic equilibrium, the aggregates can be achiral or chiral (racemic mixture). When achiral aggregates are placed in a chiral hydrodynamic flow, depending on their size, shape, and flexibility, deformation can occur leading to a selected chiral state. When a racemic mixture is subject to the same chiral bias, enantiomeric separation is possible. Indeed, mechanical chiral resolution has been recognized as a new and promising method to separate chiral objects at different scales, from macro down to near molecular level.[51] Going to supramolecular systems that are far from equilibrium, the actual coupling between the chiral perturbation and the growing aggregates again depends on the same parameters and could exert its influence at various stages along the kinetic pathway. The complexity of these processes, together with the experimental problems related to detection, make it rather difficult to establish an exact mechanism. However, if the basic building blocks of the supramolecular structure contain one or more chromophoric centers, a plethora of spectroscopic techniques are available to follow the kinetics of growth at different time scales. Combining linear and CD, or, even better, exploiting the state-of-art method of Mueller Matrix spectroscopy is a valuable tool to shed light in this kind of investigations.[50]

2.2. *J-aggregates of water-soluble porphyrins: Good spectroscopic probes for chirality detection*

Water-soluble sulfonated porphyrins have been largely investigated for their propensity to self-aggregate. In particular, the meso-phenyl substituted porphyrins bearing up to four sulfonic groups on the *para* position of the phenyl moieties have been the most studied (TPPS$_x$, x = 1 − 4 is the number of sulphonic groups, see Figure 7).[52–56]

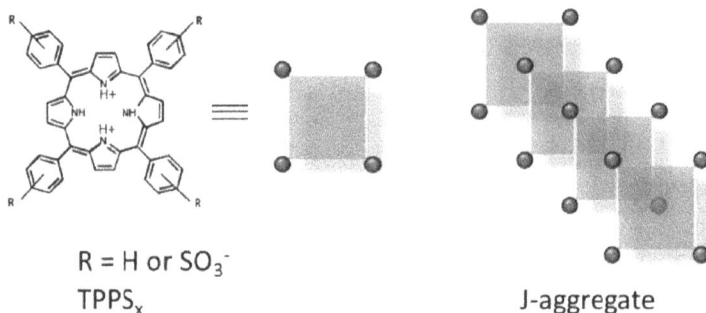

R = H or SO$_3^-$

TPPS$_x$ J-aggregate

Figure 7.   Basic structure and sketch of meso-sultanate day porphyrins (*left*, TPPS$_x$, where x = 1 − 4 is the number of sulfonate groups in para position of the phenyl groups). J-aggregates are formed by the edge-to-edge arrangement of the diacid monomers mainly through electrostatic interactions (*right*).

   When the inner nitrogen core of these molecules is protonated, under acidic conditions and/or in the presence of cationic species, the interplay of electrostatic, hydrogen bonding, and π-interactions leads to the formation of J-aggregates. These supramolecular species are characterized by a strong red-shifted absorption J-band, together with a blue-shifted H-band, originating from exciton coupling of the electronic transitions in the starting monomeric units (Figure 8).[57] Depending on the experimental conditions (pH, ionic strength, porphyrin concentration, temperature) and the mixing protocols, a variety of nanoarchitectures are observed, including nanotubes and nanoribbons.[58,59] As a consequence, even slight changes in the selected initial conditions (e.g., a specific salt used to control the ionic strength) deeply influence the kinetic rates and pathway to aggregation, determining structural modifications.[60]

   Due to its amphiphilic character, TPPS$_3$ forms prevalently quite flexible nanoribbons, while the highly charged TPPS$_4$ gives more rigid nanotubes. Chirality can be induced in these aggregates by the use of chemical templates, both simple molecules (e.g., tartrate or aminoacids)[61,62] or polymers (e.g., polylysine, Deoxyribonucleic acid (DNA)),[63–66] and by physical chiral bias (e.g., stirring).[36–38] The relevant spectroscopic features (strong absorption in the visible range) make these supramolecular assemblies quite interesting as model systems, since chirality can be easily detected by CD spectroscopy. In the case of electronically coupled chromophores, exciton split CD spectra are observed with typical bisegnated bands that can be correlated to the handedness of the system (Figure 9).[67]

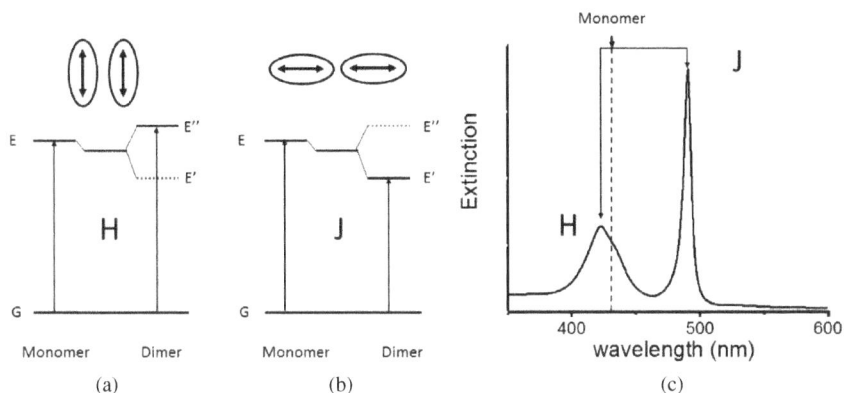

Figure 8.   (a) Exciton coupling according to Kasha *et al.*[57] in a face-to-face arrangement of transition moment dipoles (H-type coupling, blue shift), (b) exciton coupling in an edge-to-edge arrangement of transition moment dipoles (J-type coupling, red shift), and (c) typical extinction spectrum of TPPS$_4$ J-aggregates in the B-band region, showing the residual monomeric diacid porphyrin at 434 nm (shoulder), the H-band at 420 nm, and the J-band at 490 nm.

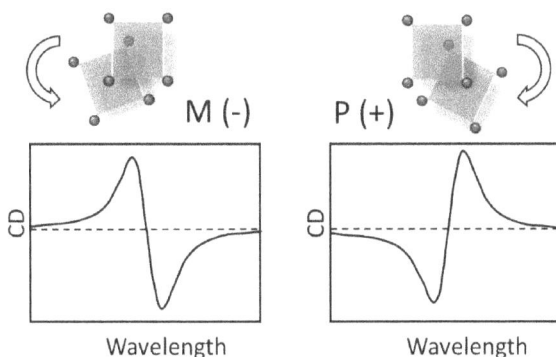

Figure 9.   Porphyrin aggregates with a left (right)-handed arrangement (M(−) or P(+) helicity) of the chromophores (*left and right*) and their corresponding bisegnated negative (positive) CD exciton spectra.

2.3. *Some initial experiments on stirring*

In 1976, Honda and Hada reported that the sign of CD spectra of J-aggregates formed in solution by 1,1'-diethyl-2,2'-cyanine chloride, a pseudo-cyanine (PIC) dye, when obtained under stirring by a magnetic bar, were related to the sense of rotation.[68] Various other research groups attempted unsuccessfully to reproduce these experimental results. Therefore, this paper was strongly criticized, and most of the arguments were based on the contribution of linear dichroism as strong artifact in the experimental CD spectra,[69] even if the actual knowledge on these systems has somehow confirmed the original interpretations of the chiroptical response.[70] Indeed, J-aggregates of this dye and related pseudo-cyanines are prone to form quite long fibers that can be easily aligned by flow. The first experimental evidence of optical activity in a supramolecular system based on porphyrins was provided later in 1993 by Ohno *et al.* in a report on the spectroscopic characterization of J-aggregates of the diacid form of the water-soluble porphyrin, tetrakis-*meso*(4-sulfonatophenyl)porphyrin (TPPS$_4$).[71] Together with the detection of CD signal in aggregates prepared in the presence of various enantiomeric form of tartaric acid, they described the observation of opposite CD spectral features for samples prepared by stirring the solutions clock- or counterclockwise. Also in this case, the scarce reproducibility of the results led to a weak impact on the scientific community.

2.4. *Effect induced by vortex flow during the aggregation of porphyrin mesophases*

In 2001, Ribò and co-workers again boosted the interest toward the impact of stirring on the supramolecular chirality, showing that indeed it was possible to obtain a statistical relationship between the CD sign and the rotation sense imposed to the solutions during the formation of J-aggregates of the diacid form of 5-(phenyl)-10,15,20-(4-sulfonatophenyl) porphyrin (TPPS$_3$).[36] Protonation of the macrocycle nitrogen atoms lowers the total negative charge of this molecule, while increasing the ionic strength affords a further screening effect, thus lowering the repulsive electrostatic barrier to the approach among the single units. This initial step is at the basis of the formation of a J-type aggregate, where the single porphyrins self-organize into linear assemblies mainly stabilized by the ion-pairing of the positively charged nitrogen core and the negatively charged sulfonate moieties of the peripheral substituent groups. Thin and easily deformable nanoribbons are mostly formed under such conditions. The preparation of the samples was achieved by slow concentration of acidified and salted diluted porphyrin solutions using two rotovapor

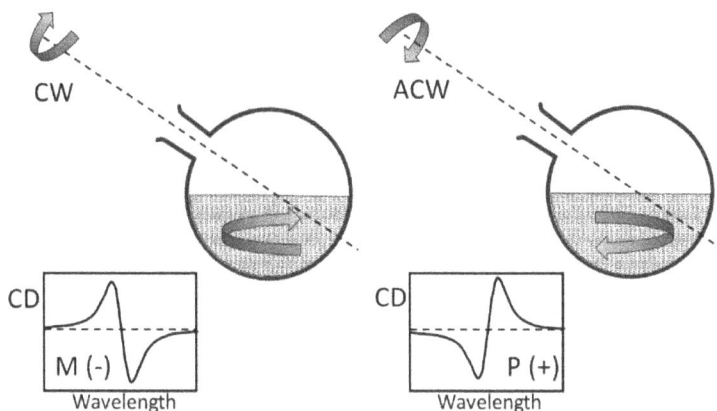

Figure 10. Schematic outcome of the Ribò's experiment on TPPS$_3$ J-aggregates. Acidified diluted solutions of TPPS$_3$ are concentrated in a rotary evaporator and aggregates are slowly formed upon increasing porphyrin, salt, and acid concentration. CW (ACW) stirring during the aggregation leads to negative (positive) CD spectra.

apparatus rotating at a speed of ca. 600 rpm and operating in opposite rotation sense. In order to make the evaporation process efficient, temperature was set to 55°C, reaching the final volume and concentration of reagents in about two hours. CD spectra measured on these samples evidenced a good statistical correlation between the observed sign of the exciton couplet and the rotation sense (ca 86%), being positive for an anticlockwise (ACW) rotation (Figure 10).

The measured CD spectra remained unaltered for several months, pointing to an irreversible effect. Blank experiments were conducted without using the previous rotovapor apparatus, on unstirred samples prepared adding concentrated acid to a salted porphyrin solution at the same final concentration of the rotated ones and at room temperature. In this case, CD spectra analysis shows a random distribution of signs (ca. 50%), in line with a simple symmetry breaking process. A general outcome of these experimental findings is that the scenario is different with respect to what is observed in symmetry breaking during crystallization. In this latter case, despite very high enantiomeric excess (*ee*) there is no correlation between the obtained enantiomer and the stirring sense. In TPPS$_3$ experiments, the process is strongly biased by the perturbation, even if it is difficult to assess the *ee*. The mechanism proposed for the enantioselection was based on the diastereotopic trajectories of small oligomers in the chiral hydrodynamic flow of the vortex with respect to longer H-type aggregates. The initial chiral fluctuations imposed by the two possible arrangements of the porphyrin units (P or M) could be further amplified and propagated by an autocatalytic process during a diffusion-controlled aggregation growth. Quite recent models on the hydrodynamic flow inside a rotovapor apparatus have successfully described these experimental results.[72]

In later investigations, this effect was tested on the complete series of TPPS$_x$ porphyrins, but a statistical correlation with the vortex direction was clearly identified only for TPPS$_3$.[73] AFM images on freshly prepared J-aggregates of TPPS$_3$ revealed their straight ribbon-like structure, whose thickness points to a double-layered arrangement of porphyrins. When solutions are aged, some degree of folding is evident, and in stagnant and not rotated samples, aggregates become larger irregularly folded and appear as bundles. Only when solutions are stirred are well-defined spirals are detected. Furthermore, the protocol for preparing the supramolecular aggregates was changed, using a constant porphyrin concentration and stirring the solution inside a cylindrical tube with a magnetic bar. In this case, the sign of CD signals is inverted with respect to rotary-evaporated samples as an indication of the flow at the wall as the main responsible for the effect.[74] A detailed analysis of the hydrodynamic flow in magnetic stirred samples has pointed out a strong dependence on the geometrical shape of the vessel and a clear inversion of the chirality sign between the central descending and the lateral ascending flows.[75] On these bases, flexible spirals possessing chirality follow different diastereotopic trajectories in an already chiral flow. The disfavored ones are driven toward a region of chaotic flow that, by friction, disassembles them in smaller oligomers. Even if this is a plausible hypothesis, another simple hydrodynamic model has pointed out the role of the solid–liquid interface in nucleation and seeding the chiral aggregates.[76] The hydrodynamic torque acting on a growing nanoribbon free in solution affords almost exclusively its alignment. Only if the nanoribbon is anchored somehow to a solid interface can twisting occur and chirality appear.

2.5. *Vortexing and chiral separation*

What happen if a racemic mixture of supramolecular aggregates is subject to a vortex flow? In 2010, Purrello and co-workers showed that a solution could be enantiomerically enriched by selectively depositing one

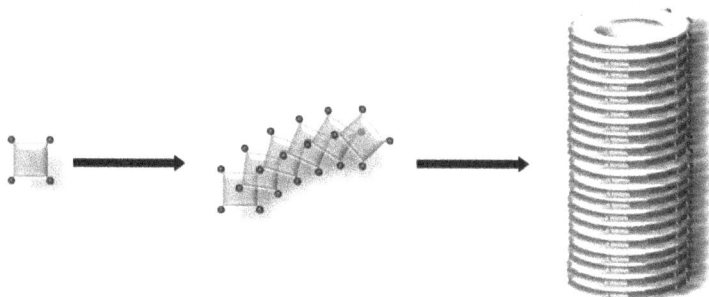

Figure 11.   Under the proper experimental conditions, the diacid form of TPPS$_4$ porphyrin forms nanotubes, due to the helicoidal stacking of the monomers.

of the two enantiomorphous J-aggregates on the walls of the reaction vessel.[38] These experiments were performed using the TPPS$_4$ porphyrin. One difference with TPPS$_3$, the further negatively charged sulfonate group present at the periphery on this compound removes the amphiphilic character and induces a structural change under most of the experimental conditions toward a nanotube (Figure 11).[77]

Despite the achiral initial structure of this porphyrin, aggregation under acidic conditions affords solutions that exhibit strong CD signals, even in the absence of any apparent chiral bias. Different hypotheses on the origin of this optical activity have been proposed: (i) the presence of adventitious tiny traces of chiral contaminants always present even in ultrapure water, capable of inducing the chirality in these achiral porphyrin J-aggregates,[78] (ii) the nanoarchitecture being intrinsically chiral and a genuine symmetry breaking process occurring, or (iii) the nanoarchitecture being intrinsically chiral and the adventitious presence of chiral contaminants leading to a scalemic mixture. Actually, the most accredited hypothesis is the last one, even if a definitive answer is still lacking.

Another important point is that, different from the monomeric porphyrin, TPPS$_4$ J-aggregates have a strong tendency to stick to the wall of the quartz cuvette, where they retain their CD features. Initially, it is worth noting that CD spectra measured *in situ* under stirring are more intense that the resting solutions and the sign of the observed couplet depends on the rotation sense, being negative for clockwise (CW) and positive for ACW rotation. Furthermore, if the stirrer is stopped, the CD signals are restored to their initial aspect (Figure 12). On the nature of this phenomenon, we will discuss later on this paragraph.

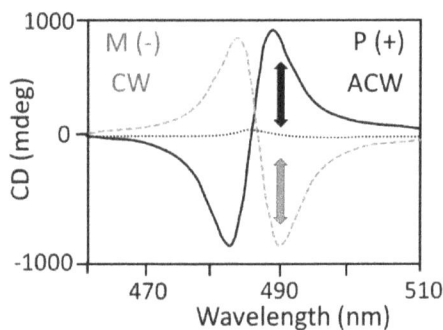

Figure 12.   CD reversible spectra obtained on samples containing TPPS$_4$ J-aggregates under magnetic stirring (CW, dashed line; ACW, continuous line). The intensity of the spectra increases with the stirring speed and turns to the initial small value on stopping the stirrer. The dotted line is the CD spectrum of the resting solution.

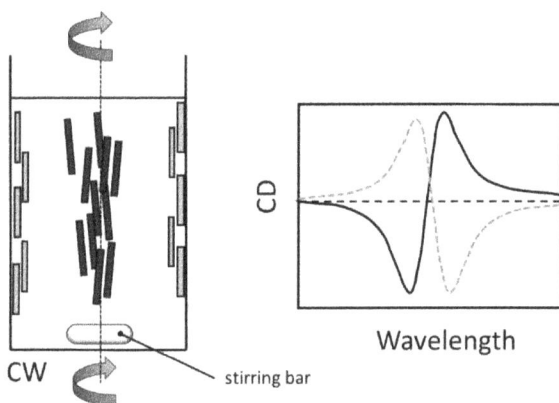

Figure 13.   Basic description of the enantioseparation experiment on TPPS$_4$ J-aggregated samples. Under CW stirring for 24 h, the solution is enriched by the P (+) (solid black line) enantiomorphous aggregate, while the M(−) one (dashed grey line) is deposited on the cell wall. When an ACW motion is applied, the opposite trend is observed.

The basic experiments revealed that applying CW rotation through a magnetic stirring bar for 24 hours the (−)-M J-aggregate is deposited on the wall of the cuvette, while ACW rotation leads to a deposition of the opposite (+)-P J-aggregate (Figure 13). These experimental findings hold either starting from solutions enriched by a specific enantiomeric aggregate or from racemic mixtures. In this latter case, the CD spectrum of the

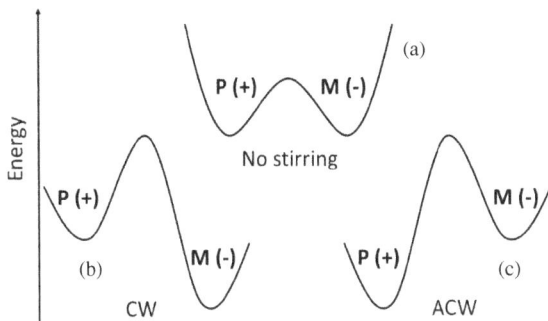

Figure 14.   Schematic representation of the energy profiles suggested in D'Urso *et al.*[38] for the effect of stirring on a racemic mixture of porphyrin J-aggregates: (a) in a standing solution the two enantiomorphous J-aggregates have the same energy, (b) under CW stirring the M(−) form is more stable, and (c) under ACW stirring the P(+) enantiomer is the preferred one.

starting sample is obviously silent. After stirring in a specific sense, the solution is enriched by one of the enantiomers, while the other is deposited on the solid surface.

Even if the statistical analysis of the results was not so strong, a clear indication that stirring is able to operate an enantioselective deposition of aggregated species has been pointed out. The proposed explanation for this effect is based (i) on the ability of the vortex flow to shift the thermodynamic equilibrium between the two enantiomorphous aggregated species (Figure 14) and (ii) the observation that the major species in solution usually sticks onto the solid surface.

Chemical chiral bias can compete with this physical bias and if a chiral *tris*-chelated cationic metal complex is added to the J-aggregates, two different scenarios occur: (i) if the chiral metal complex is at high concentration, it competes efficiently with the vortex action and the deposited species is the predominant one in solution, irrespective of the sense of the vortex and (ii) if the chiral complex is at very low concentration, the vortex action prevails on the chemical chiral bias and the deposited sample displays chirality correlated with the rotation sense.

The competition between chemical and vortexing chiral bias has been further evidenced in other aggregated systems based on surfactants and their fine balance can afford a precise control over the chirality, achieving in some case a net compensation producing zero effect.[79]

The observation of temporary strong CD signals for stirred samples raises the question about instrumental artifacts, due to the use of standard spectropolarimeter. As outlined above, Mueller Matrix spectrometry is the current state-of-art technique to overcome this problem. In the experiments described in this paragraph, strong CD signals are detectable and their sign depends on the vortex sense. Strong chiroptical response in samples containing supramolecular species induced by convective motion and/or vortex have been reported by different groups, and the outcomes have been regarded as a potential warning about a possible misinterpretation of the effect.[80] In 2007, two independent articles from the groups of Aida[81] and Meijer[82] on different supramolecular assemblies reported very strong chiroptical effects induced by stirring. Orientation of fiber-like supramolecular structures in solution by vortexing, shaking, convective flow, or even by acoustic waves[83] leads to contribution of linear dichroism and birefringence, whose combination determines the observed chiroptical response of the sample. As a consequence, the observed CD signals are the results of a macroscopic effect, not a real molecular chirality. In the case of TPPS$_4$ J-aggregates, the temporary and reversible effects imposed by stirring could find a similar explanation, even if a detailed Mueller Matrix spectroscopical analysis on these samples revealed the presence of true CD, as a specific marker of molecular chirality.[84,85]

2.6. *Rotation and effective gravity*

The hydrodynamic torque experienced by a supramolecular species during its growth is effective in generating a structural folding or twisting if the aggregate is large enough and is anchored somehow on a solid surface. Smaller aggregates free in solution should not even be oriented due to Brownian thermal motion. *Is it possible to impose a chiral bias at the beginning of an aggregation process, when the basic seeds are still small?* In Ribò's experiment, the observed folded spiral aggregates are large enough to have experienced the hydrodynamic torque, and there was no experimental evidence on the aggregation stage really important for the chirality transmission from the chiral bias to the growing system. Furthermore, if the torque effect is acting on larger aggregates, these species could also experience barodiffusion, i.e., the diffusion due to gravity. Indeed, gravity plays an important role during crystallization, and its substantial suppression (microgravity conditions) leads to minimization of convective motions.[86] Also, self-assembling of supramolecular systems controlled by diffusion, e.g., microtubules, is largely hampered by gravity reduction.[87] On these bases, one could ask if gravity is really effective in controlling the chirality in supramolecular systems. To shed light on this issue, an experimental approach was presented in the literature using the $TPPS_3$ porphyrin under aggregating conditions similar to those exploited in Ribò's experiments at constant porphyrin concentration, but with a chiral bias applied only for a short time at the beginning of the aggregation process.[41] Accordingly, the kinetic conditions were chosen in order to reach complete aggregation after three days. In order to create a controlled gravity, the experiments were performed inside a strong magnet and rotation has been applied simultaneously for a time period quite short (up to 120 min) with respect to the overall aggregation time. The basic experimental set-up is shown in Figure 15(a).

A tube holder filled with a number of vials containing the same initial aggregating solution is located in the internal bore of a large magnet (16–25 T) and rotated CW or ACW at constant temperature. The magnetic levitation force to which a diamagnetic sample is subject inside the bore is proportional to the product of the magnetic field strength and the magnetic field gradient and is position-dependent (Figure 15(b)). The magnetic levitation force, changing its sign with the quote, affords an easy way to control the effective gravity exerted on the samples. As a result: (i) for a vial placed at the magnet center, the magnetic field is at maximum intensity but magnetic force is null (normal gravity, G_n), (ii) for a vial below the center, the effective gravity is concordat to the terrestrial one but enhanced (positive), (iii) for a vial above the center, the effective gravity is inverted (negative), and (iv) in this particular set-up, the vial at the top is close to null gravity conditions (levitation). After being exposed to the chiral bias, the vials were extracted from the magnet, placed on a bench, and CD spectra were measured on fully aggregated samples after three days. The basic results from these experiments are displayed in Figure 16. In the absence of magnetic field, there is no chiral selection and the measured CD intensity is quite low. When the magnetic field is applied, CW rotation gave negative CD couplets for samples under positive effective gravity and positive CD couplets for samples under negative effective gravity. Switching rotation to ACW, all the results were inverted.

A statistical analysis on the experiments revealed an enantioselection close to 100%. The outcomes of this investigation are manifold. First, it is possible to combine magnetic field and rotation to achieve

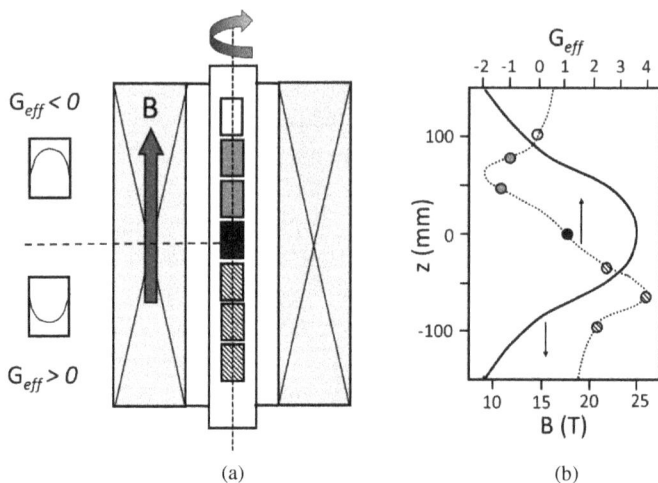

Figure 15.   (a) Experimental set-up for the experiments reported in Micali *et al.*[41] Seven vials containing aliquots of the same aggregating solution of $TPPS_3$ porphyrin were placed in a rotating tube (CW or ACW, 15 Hz speed), inside the bore of a magnet (16 T or 25 T). Magnetic field and rotation were applied to the samples at the beginning of the aggregation process for a variable period (15–120 min). The sample at the center of the magnet (black) experiences the maximum field and an effective gravity close to the normal gravity. The three sample below the center (dashed) are subjected to enhanced effective gravity (the meniscus is downward, $G_{eff} > 0$), while the two above the center (grey) experience an inverted effective gravity (the meniscus is upward, $G_{eff} < 0$). The sample at the top of the tube is near levitation (white), and (b) plot of the magnetic field intensity (solid black line) and the effective gravity G_{eff} (dotted line) as a function of the position z inside the magnet.

enantioselection in a supramolecular system. Actually, the magnetic field plays a double role: (i) it controls the effective gravity, by tuning the levitation forces and (ii) it aligns the nanoaggregates.

   Alignment is then a prerequisite for nanoribbons to start their folding under the hydrodynamic flow generated by the rotation. Second, the chiral physical bias can be easily applied and removed from the samples without any further manipulation. Third, a falsely chiral bias, i.e., the combination of rotation and gravity, is able to induce chirality into an aggregating supramolecular system (Figure 17). This finding is an experimental confirmation of the theoretical prediction by L. Barron that, under such influences, only if a system is kinetically controlled and far from equilibrium can it be driven toward a specific enantiomeric state.[88,89] Fourth, a short perturbation at the very beginning of an assembling process is sufficient

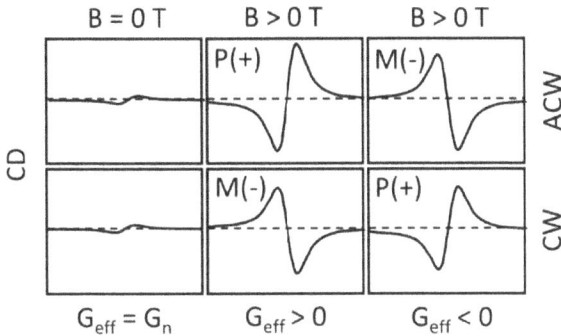

Figure 16. Typical CD spectra for the aggregation of $TPPS_3$ evidencing the effect of rotation (ACW, upper; CW, lower) and magnetic field on chiral selection. When the magnetic field is turned off (left panels), CD intensity is low, the effective gravity is equal to the normal gravity, rotation is ineffective, and no selection is observed. When the magnetic field is turned on (middle and right panels), ACW rotation leads to P(+) enantiomer in the samples with positive effective gravity and to M(−) enantiomer in samples with negative effective gravity. The reverse situation holds switching to CW rotation.

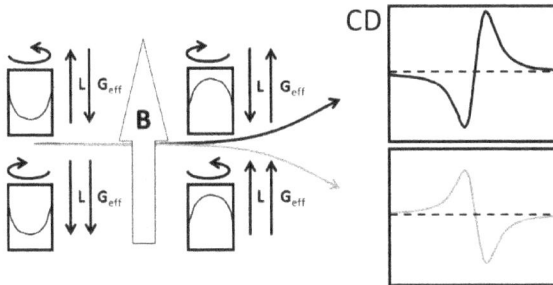

Figure 17. The various possible combinations of the angular momentum **L** and effective gravity $\mathbf{G}_{eff}$ and their relation with the observed CD signals: an antiparallel arrangement of both vectors leads to the P(+) enantiomer (solid black line), while the parallel arrangement of the vectors gives the M(−) enantiomer (solid grey line). The magnetic field **B** is showed to outline its role in aligning the nanoribbons.

to promote enantioselection. Figure 18 shows the time evolution of the extinction measured for the band corresponding to the formation of J-aggregates. It is quite evident that the early stage of this kinetic process, corresponding to nucleation, is the period where the chiral bias has been applied. During this initial period, particle size measurements have

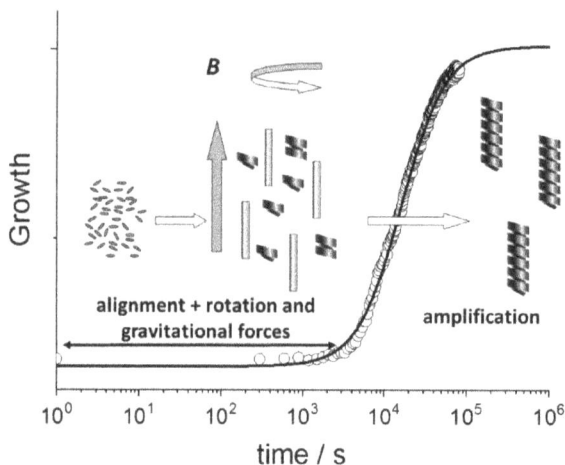

Figure 18.   Proposed mechanism for the chiral selection in TPPS$_3$ nanoaggregates under the bias of rotational and gravitational forces. The chiral physical bias is applied to the samples only at the beginning of the aggregation process, when the system is at its early nucleation phase. During this stage, short nanoribbons form, are aligned by the magnetic field, and are twisted into short nanohelices. After a couple of days, chirality is amplified and longer helices develop. The solid black line is the best fitting to the extinction data corresponding to the kinetic growth of the aggregates in solution.

demonstrated that the aggregates are even smaller than 40 nm. Indeed, this experimental finding provides some hints for a fundamental question about the spontaneous symmetry breaking and its implications in the origin of homochirality in life. Actually, spontaneous symmetry breaking is quite common in nature, and many examples can be found during crystallization.[4,42,43] The real question for the total selection of only one of the two possible enantiomers in a prebiotic scenario is related to further mechanistic pathways explaining how chirality could propagate and amplify. The experimental approach here described shows that starting from an initial chiral burst, the pathway to complexity finds a way to converge somehow into a selected chiral state.

2.7. *Vortexing in microfluidic devices*

In the previous sections, the importance of the protocol applied to prepare a supramolecular system has been outlined. Even if care could be taken,

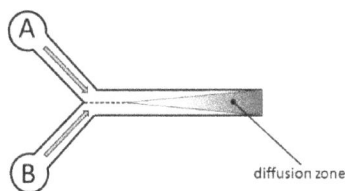

Figure 19.   Schematic sketch of a microfluidic device operating in continuous flow mode in which two liquid solution (A and B) are under a laminar flow regime.

the macroscopic gradients generated by mixing solutions could be responsible for different or irreproducible results. A methodology that allows a precise control of mixing, diffusion, and mass transport is based on microfluidic devices.[90] These latter have sub-millimeter sizes and permit fine tuning of many factors, operating in a laminar flow regime, affording a high level of reliability and reproducibility. A proper design of a microfluidic device can include more inlet channels to inject different solutions of reagents, whose individual flow rates can be accurately controlled. Solutions in the same solvent can co-flow in the microchannel and mix only by diffusion at the interface, without turbulence, with a contact time that is a function of the flow rate and the length of the channel (Figure 19). Such favorable conditions have been largely employed for growing crystals.[91] In the case of supramolecular systems, this technique has provided a way to achieve a better kinetic control of the assembling process.[92]

TPPS$_4$ porphyrin is able to form chiral supramolecular systems if its aggregation is triggered in the presence of proper chiral species. Indeed, the spontaneous symmetry breaking observed for this molecule in the apparent absence of any chiral bias has been ascribed to the presence of trace chiral pollutants.[78] Mixing aqueous solutions of TPPS$_4$, under acidic conditions, with a chiral surfactant in a microfluidic channel has proven that transfer of chirality occurs in this system within tens of milliseconds, that is the residence time of the mixed solution inside the device.[93] In a quite recent report, M. Liu and coll. used this approach to demonstrate the possibility to control the chirality of TPPS$_4$ J-aggregated nanotubes and of supramolecular gels formed by a C$_3$-derivative of benzene-1,3,5-tricarboxamide.[37] The particular design of the microfluidic device allows to mix the reagents and drive them into a series of successive inclined chambers located on both sides along the path and able to generate counter-rotating microvortices (Figure 20). The liquid flowing on one side is forced into the chambers, where it experiences the same kind of rotation

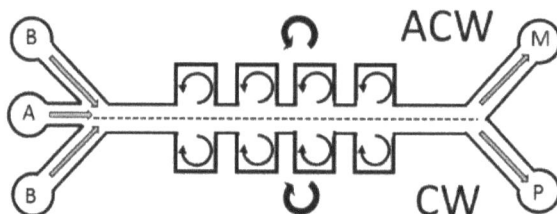

Figure 20.   Schematic representation of the experimental microfluidic device used in Sun *et al.*[37] to select chirality during aggregation. The reagents were introduced through the inlets A and B and the mixed samples were collected through the outlet M and P. A series of consecutive chambers (ten in the actual device) are built on both sides of the central channel, specifically designed to generate ACW or CW vortex flows.

and is collected in a separate outlet after a residence time of few milliseconds. Simulation of the flow inside the device has provided an accurate description of the vortex inside the microchamber, where the fluid can follow CW or ACW circular trajectories, generating P or M helicity, with speed in the order of 10^4 rpm. The samples collected at the two different outlets, after equilibration, show CD spectra whose sign correlates with the rotation sense (almost 100%).

Experiments on the supramolecular gelification displayed analogous results in terms of enantioselection. In both cases, the aggregation time was estimated shorter that the residence time inside the microfluidic channel. Apart from the potential application of this technique to control the chirality in supramolecular species, further implications on homochirality could be envisaged. An interesting hypothesis by these authors identifies the microporous rocks close to the hydrothermal vent in the oceans as possible sites for enantioselection. The high pressure of the fluid flowing inside such micropores, together with other extreme conditions, could have generated asymmetric synthesis in a prebiotic scenario.

3. Thermophoretic Forces

As pointed above, when long fiber-like supramolecular aggregates are present in solution, they can be aligned by simple shaking, stirring, or by thermal convection. In particular, Meijer and coll. reported unusual CD bands in solutions of chiral and achiral oligo(*p*-phenylene vinylene) derivatives.[82] In addition to macroscopic CD signals originating from combination of linear dichroism and linear birefringence due to alignment during vortexing, they showed also that artifactual CD was due to linear dichroism caused by alignment originating from convective flow. Indeed, the compounds used in these investigations are monomeric in solution at high temperature, while they are able to assemble into long fibers upon cooling. The length of the fibers appeared to be important in detecting the chiroptical response from the samples. Also, alignment is partially reduced on decreasing the path length of the optical cell, thus largely influencing the results.

More recently, the induction of chirality in supramolecular systems by thermophoresis was investigated by exploiting porphyrin-based aggregates. As described in the previous sections, $TPPS_x$ porphyrins are able to self-assemble into a large variety of structures, mainly stabilized through electrostatic interactions, that confer a substantial rigidity to these nanoarchitectures. The proper choice of an uncharged porphyrin derivative bearing four flexible polyethyleneglycol chains on the peripheral phenyl meso groups allows to obtain aggregates where the constituent units interact weakly through an interplay of hydrogen bonding and hydrophobic interactions (stacking interactions). This compound is soluble as a monomer in organic solvents, while it extensively aggregates in water where fractal clusters were detected by light scattering techniques.[44] When solutions of these aggregates were placed into the conventional thermostat inside the spectropolarimeter compartment, bisegnated CD spectra were detected whose intensity and sign are related to the temperature imposed to the device. Using a home-made thermostat apparatus with a higher accuracy ($\pm 0.05°C$), the intensity of the observed bands was smaller and decreased on decreasing the difference with the room temperature. If a precision two-stage thermostat was exploited ($\pm 0.01°C$), the CD spectra were silent. These experimental findings were interpreted as a chiral motion of the aggregates and not of the solvent, imposed by thermophoresis. Further investigations revealed that CD spectra display opposite sign if they undergo cooling or heating treatments. Also, linear dichroism (LD) contributions were ruled out by independent measurements on these samples.

A model of the thermal flow inside the cuvette filled with water was performed using a finite element approach, showing that: (i) when all the six walls of the vessel have the same temperature, a chaotic flow is observed and (ii) when an asymmetric distribution of temperature on the walls is simulated, a flow from the hotter to the colder zone arises, changing its sense on inverting the imposed pattern of temperature (Figure 21).[45]

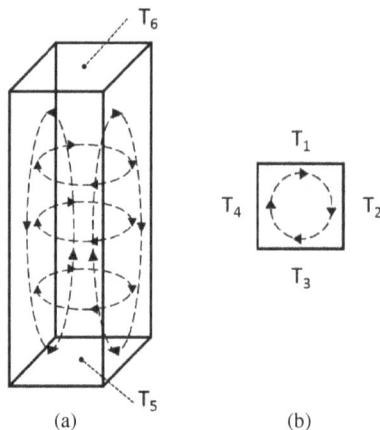

Figure 21.   Schematic sketch of the convective flow inside a cell with an asymmetric small temperature difference among the walls ($T_1 > T_2 > T_3 > T_4$; $T_5 < T_6$). In the case of a reverse distribution, the internal convective flow inverts.

The flexibility and deformability of these aggregates also lead to an easy control of their chiroptical response through the application of gentle rotation (by hands) of the vessel or vortexing. In this latter case, the CD signals are much larger and become very intense on increasing the rotation speed, changing also the sign on inverting the rotation sense.

This particular phenomenon has been also reported for other structural motifs, such as nanosheets formed by a charged achiral pyrene trimer.[46] These 2D structures can be prepared by thermal annealing, and their CD and LD spectra are strongly affected by the preparation protocol, that in turn determines the size of the resulting aggregates. In the case of an isothermal aggregation, the nanosheets are small and the samples are CD and LD silent. When a slow thermal annealing is performed, larger aggregates are present in solutions and they can be aligned by thermophoretic forces and the supramolecular chirality can be controlled by the hydrodynamic flow imposed through a gentle vortex.

4. Concluding Remarks

Since the separation of optical isomers, Louis Pasteur tried to relate the presence of chiral compounds with physical forces in nature.[94] A static magnetic field, or subsequently the combination of an electric and a magnetic field, as suggested by Pierre Curie,[95] were erroneously proposed as potential chiral forces. Later, Lord Kelvin recognized that *the magnetic rotation has neither a right-handed nor left-handed quality (that is to say, no chirality)*.[1] Notwithstanding, these initial ideas fascinated many researchers, generating unfruitful efforts to obtain absolute asymmetric synthesis.[96] Anyway, Pasteur correctly recognized that rotation associated with a translational motion is a chiral bias. Nowadays, an increasing number of scientific papers report how the hydrodynamic flow of a solvent is able to control the handedness of a growing supramolecular system or to efficiently deform flexible aggregates from the nano- up to the microscale. The flow can be mechanically generated by stirring a fluid or even by slight thermal gradients and the complexity of the problem has caused some initial debate on the reproducibility of the results. Actually, theoretical models afford a reliable description of the fluid motion, thus explaining the emergence of chirality under such conditions. Also, a longstanding question about the possibility that a falsely chiral bias (e.g., rotation and gravity) could induce absolute enantioselection when acting on a reacting system far from thermodynamic equilibrium has been solved.[89]

The possibility to use a series of physical bias to achieve absolute asymmetric synthesis could open new potential applications in the synthesis of many pharmaceutical chiral compounds. In this context, the possibility to operate an asymmetric synthesis without the need of further purification or separation steps could represent a huge improvement. Recently, Ribò and coll. have reported an asymmetric synthesis mediated by chiral supramolecular assemblies, whose handedness is controlled by the hydrodynamic flow.[97] Even if this first example is still far from real application, it constitutes the start of a new way to introduce chirality in a chemical reaction. From a fundamental point of view, the homochirality observed in life has received large attention. In the plethora of mechanisms proposed for the emergence of a specific handedness in many essential organic compounds, the forces discussed in this chapter could be plausible, as many other explanations. Symmetry breaking is a rather common phenomenon, but a specific bias is needed to get enantioselection.[4] Growing evidence is suggesting that chiral hydrodynamic flow and thermal convection can efficiently imprint chirality to a supramolecular system under formation, even if acting just in the early stage of aggregation. Propagation, amplification, and self-healing mechanisms are needed to explain the eventual formation of only one enantiomer. Whether these asymmetric forces could be responsible for the chiral selection under prebiotic conditions still remains an open question.

References

1. L. Kelvin, *Baltimore Lectures on Molecular Dynamics and the Wave Theory of Light*, C. J. Clay & Sons, London, UK (1901).
2. D. B. Amabilino, Chiral nanoscale systems: Preparation, structure, properties and function, *Chem. Soc. Rev.* **38**(3), 669–670 (2009).
3. M. Liu, L. Zhang, and T. Wang, Supramolecular chirality in self-assembled systems, *Chem. Rev.* **115**(15), 7304–7397 (2015).
4. P. L. Luisi, *Emergence of Life: From Chemical Origins to Synthetic Biology*, Cambridge University Press, Cambridge (2006).
5. G. H. Wagnière, *On Chirality and the Universal Asymmetry*, Wiley-VCH, Weinheim, Germany (2007).
6. A. Guijarro and M. Yus, *The Origin of Chirality in the Molecules of Life*, Royal Society of Chemistry, Cambridge, UK (2009).
7. G. Balavoine, A. Moradpour, and H. B. Kagan, Preparation of chiral compounds with high optical purity by irradiation with circularly polarized light, a model reaction for the prebiotic generation of optical activity, *J. Am. Chem. Soc.* **96**(16), 5152–5158 (1974).
8. J. J. Flores, W. A. Bonner, and G. A. Massey, Asymmetric photolysis of (RS)-leucine with circularly polarized ultraviolet light, *J. Am. Chem. Soc.* **99**(11), 3622–3625 (1977).
9. J. Bailey, A. Chrysostomou, J. H. Hough, T. M. Gledhill, A. McCall, S. Clark, F. Ménard, and M. Tamura, Circular polarization in star-formation regions: Implications for biomolecular homochirality, *Science* **281**(5377), 672–674 (1998).
10. V. Aquilanti and G. S. Maciel, Observed molecular alignment in gaseous streams and possible chiral effects in vortices and in surface scattering, *Orig. Life Evol. Biosph.* **36**(5–6), 435–441 (2006).
11. T. D. Lee and C. N. Yang, Question of parity conservation in weak interactions, *Phys. Rev.* **104**, 254 (1956).
12. C. S. Wu, E. Ambler, R. W. Hayward, D. D. Hoppes, and R. P. Hudson, Experimental test of parity conservation in beta decay, *Phys. Rev.* 105, 1413 (1957).
13. A. R. A. Palmans and E. W. Meijer, Amplification of chirality in dynamic supramolecular aggregates, *Angew. Chem. Int. Ed.* **46**(47), 8948–8968 (2007).
14. R. Randazzo, A. Mammana, A. D'Urso, R. Lauceri, and R. Purrello, Reversible "chiral memory" in ruthenium tris(phenanthroline)- anionic porphyrin complexes, *Angew. Chem. Int. Ed.* **47**(51), 9879–9882 (2008).
15. A. Mammana, A. D'Urso, R. Lauceri, and R. Purrello, Switching off and on the supramolecular chiral memory in porphyrin assemblies, *J. Am. Chem. Soc.* **129**(26), 8062–8063 (2007).

16. R. Lauceri, A. Raudino, L. Monsù Scolaro, N. Micali, and R. Purrello, From achiral porphyrins to template-imprinted chiral aggregates and further. Self-Replication of chiral memory from scratch, *J. Am. Chem. Soc.* **124**(6), 894–895 (2002).

17. H. Onouchi, T. Miyagawa, K. Morino, and E. Yashima, Assisted formation of chiral porphyrin homoaggregates by an induced helical poly(phenylacetylene) template and their chiral memory, *Angew. Chem. Int. Ed.* **45**(15), 2381–2384 (2006).

18. L. Zeng, Y. He, Z. Dai, J. Wang, Q. Cao, and Y. Zhang, Chiral induction, memory, and amplification in porphyrin homoaggregates based on electrostatic interactions, *ChemPhysChem* **10**(6), 954–962 (2009).

19. R. Purrello, Lasting chiral memory, *Nat. Mater.* **2**(4), 216–217 (2003).

20. A. Sorrenti, J. Leira-Iglesias, A. J. Markvoort, T. F. A. de Greef, and T. M. Hermans, Non-equilibrium supramolecular polymerization, *Chem. Soc. Rev.* **46**(18), 5476–5490 (2017).

21. F. Marinelli, A. Sorrenti, V. Corvaglia, V. Leone, and G. Mancini, Molecular description of the propagation of chirality from molecules to complex systems: Different mechanisms controlled by hydrophobic interactions, *Chem. Eur. J.* **18**(46), 14680–14688 (2012).

22. R. F. Pasternack, A. Giannetto, P. Pagano, and E. J. Gibbs, Self-assembly of porphyrins on nucleic acids and polypeptides, *J. Am. Chem. Soc.* **113**(20), 7799–7800 (1991).

23. R. F. Pasternack, C. Bustamante, P. J. Collings, A. Giannetto, and E. J. Gibbs, Porphyrin assemblies on DNA as studied by a resonance light-scattering technique, *J. Am. Chem. Soc.* **115**(13), 5393–5399 (1993).

24. E. Bellacchio, R. Lauceri, S. Gurrieri, L. M. Scolaro, A. Romeo, and R. Purrello, Template-imprinted chiral porphyrin aggregates, *J. Am. Chem. Soc.* **120**(47), 12353–12354 (1998).

25. R. Purrello, A. Raudino, L. M. Scolaro, A. Loisi, E. Bellacchio, and R. Lauceri, Ternary porphyrin aggregates and their chiral memory, *J. Phys. Chem. B* **104**(46), 10900–10908 (2000).

26. A. S. R. Koti and N. Periasamy, Self-assembly of template-directed J-aggregates of porphyrin, *Chem. Mater.* **15**(2), 369–371 (2003).

27. L. Zhang, J. Yuan, and M. Liu, Supramolecular chirality of achiral TPPS complexed with chiral molecular films, *J. Phys. Chem. B* **107**(46), 12768–12773 (2003).

28. L. D. Barron, Symmetry and molecular chirality, *Chem. Soc. Rev.* **15**(2), 189–223 (1986).

29. W. L. Noorduin, A. A. C. Bode, M. Van Der Meijden, H. Meekes, A. F. Van Etteger, W. J. P. Van Enckevort, P. C. M. Christianen, B. Kaptein, R. M. Kellogg, T. Rasing, and E. Vlieg, Complete chiral symmetry breaking of an amino acid derivative directed by circularly polarized light, *Nat. Chem.* **1**(9), 729–732 (2009).

30. G. H. Wagnière and A. Meir, The influence of a static magnetic field on the absorption coefficient of a chiral molecule, *Chem. Phys. Lett.* **93**(1), 78–81 (1982).
31. G. L. J. A. Rikken and E. Raupach, Observation of magneto-chiral dichroism, *Nature* **390**(6659), 493–494 (1997).
32. G. L. J. A. Rikken and E. Raupach, Enantioselective magnetochiral photochemistry, *Nature* **405**(6789), 932–935 (2000).
33. P. G. de Gennes, Mechanical selection of chiral crystals, *Europhys. Lett.* **46**(6), 827–831 (1999).
34. M. H. C. Fu, T. R. Powers, and R. Stocker, Separation of microscale chiral objects by shear flow, *Phys. Rev. Lett.* **102**, 158103 (2009).
35. M. Kostur, M. Schindler, P. Talkner, and P. Hänggi, Chiral separation in microflows, *Phys. Rev. Lett.* **96**, 014502 (2006).
36. J. M. Ribó, J. Crusats, F. Sagués, J. M. Claret, and R. Rubires, Chiral sign induction by vortices during the formation of mesophases in stirred solutions, *Science* **292**(5524), 2063–2066 (2001).
37. J. Sun, Y. Li, F. Yan, C. Liu, Y. Sang, F. Tian, Q. Feng, P. Duan, L. Zhang, X. Shi, B. Ding, and M. Liu, Control over the emerging chirality in supramolecular gels and solutions by chiral microvortices in milliseconds, *Nat. Commun.* **9**, 2599 (2018).
38. R. R. D'Urso, L. Lo Faro, and R. Purrello, Vortexes and nanoscale chirality, *Angew. Chem. Int. Ed.* **49**(1), 108–112 (2010).
39. D. Edwards, K. Cooper, and R. C. Dougherty, Asymmetric synthesis in a confined vortex: Gravitational fields can cause asymmetric synthesis, *J. Am. Chem. Soc.* **102**(1), 381–382 (1980).
40. L. D. Barron, Reactions of chiral molecules in the presence of a time-non-invariant enantiomorphous influence: A new kinetic principle based on the breakdown of microscopic reversibility, *Chem. Phys. Lett.* **135**(1–2), 1–8 (1987).
41. N. Micali, H. Engelkamp, P. G. van Rhee, P. C. M. Christianen, L. Monsù Scolaro, and J. C. Maan, Selection of supramolecular chirality by application of rotational and magnetic forces, *Nat. Chem.* **4**(3), 201–207 (2012).
42. D. K. Kondepudi, R. J. Kaufman, and N. Singh, Chiral symmetry breaking in sodium chlorate crystallization, *Science* **250**(4983), 975–976 (1990).
43. C. Viedma, Chiral symmetry breaking during crystallization: Complete chiral purity induced by nonlinear autocatalysis and recycling, *Phys. Rev. Lett.* **94**, 065504 (2005).
44. P. Mineo, V. Villari, E. Scamporrino, and N. Micali, Supramolecular chirality induced by a weak thermal force, *Soft Matter* **10**(1), 44–47 (2014).
45. P. Mineo, V. Villari, E. Scamporrino, and N. Micali, New evidence about the spontaneous symmetry breaking: Action of an asymmetric weak heat source, *J. Phys. Chem. B* **119**(37), 12345–12353 (2015).

46. N. Micali, M. Vybornyi, P. Mineo, O. Khorev, R. Häner, and V. Villari, Hydrodynamic and thermophoretic effects on the supramolecular chirality of pyrene-derived nanosheets, *Chem. Eur. J.* **21**(26), 9505–9513 (2015).

47. J. Crusats, Z. El-Hachemi, and J. M. Ribó, Hydrodynamic effects on chiral induction, *Chem. Soc. Rev.* **39**(2), 569–577 (2010).

48. O. Arteaga, A. Canillas, J. Crusats, Z. El-Hachemi, J. Llorens, A. Sorrenti, and J. M. Ribo, Flow effects in supramolecular chirality, *Isr. J. Chem.* **51**(10), 1007–1016 (2011).

49. K. Okano and T. Yamashita, Formation of chiral environments by a mechanical induced vortex flow, *ChemPhysChem* **13**(9), 2263–2271 (2012).

50. O. Arteaga, Z. El-Hachemi, A. Canillas, and J. M. Ribó, Transmission Mueller matrix ellipsometry of chirality switching phenomena, *Thin Solid Films* **519**(9), 2617–2623 (2011).

51. V. Marichez, A. Tassoni, R. P. Cameron, S. M. Barnett, R. Eichhorn, C. Genet, and T. M. Hermans, Mechanical chiral resolution, *Soft Matter* **15**(23), 4593–4608 (2019).

52. J. M. Ribó, J. Crusats, J.-A. Farrera, and M. L. Valero, Aggregation in water solutions of tetrasodium diprotonated meso-tetrakis (4-sulfonatophenyl) porphyrin, *Chem. Commun.* **6**, 681–682 (1994).

53. R. F. Pasternack, K. F. Schaefer, and P. Hambright, Resonance light-scattering studies of porphyrin diacid aggregates, *Inorg. Chem.* **33**(9), 2062–2065 (1994).

54. D. L. Akins, H.–R. Zhu, and C. Guo, Absorption and Raman scattering by aggregated meso-tetrakis(p-sulfonatophenyl)porphine, *J. Phys. Chem.* **98**(14), 3612–3618 (1994).

55. N. C. Maiti, M. Ravikanth, S. Mazumdar, and N. Periasamy, Fluorescence dynamics of noncovalently linked porphyrin dimers, and aggregates, *J. Phys. Chem.* **99**(47), 17192–17197 (1995).

56. N. Micali, F. Mallamace, A. Romeo, R. Purrello, and L. M. Scolaro, Mesoscopic structure of meso-tetrakis(4-sulfonatophenyl)porphine J-aggregates, *J. Phys. Chem. B* **104**(25), 5897–5904 (2000).

57. M. Kasha, H. R. Rawls, and M. Ashraf El-Bayoumi, The exciton model in molecular spectroscopy, *Pure Appl. Chem.* **11**(3–4), 371–392 (1965).

58. Z. El-Hachemi, C. Escudero, F. Acosta-Reyes, M. T. Casas, V. Altoe, S. Aloni, G. Oncins, A. Sorrenti, J. Crusats, J. L. Campos, and J. M. Ribó, Structure vs. properties-chirality, optics and shapes — in amphiphilic porphyrin J-aggregates, *J. Mater. Chem. C* **1**(20), 3337–3346 (2013).

59. Z. El-Hachemi, T. S. Balaban, J. Lourdes Campos, S. Cespedes, J. Crusats, C. Escudero, C. S. Kamma-Lorger, J. Llorens, M. Malfois, G. R. Mitchell, A. P. Tojeira, and J. M. Ribó, Effect of hydrodynamic forces on meso-(4-sulfonatophenyl)-substituted porphyrin J-aggregate nanoparticles: Elasticity, plasticity and breaking, *Chem. Eur. J.* **22**(28), 9740–9749 (2016).

60. A. Romeo, M. A. Castriciano, I. Occhiuto, R. Zagami, R. F. Pasternack, and L. Monsù Scoloro, Kinetic control of chirality in porphyrin J-aggregates, *J. Am. Chem. Soc.* **136**(1), 40–43 (2014).

61. M. A. Castriciano, A. Romeo, R. Zagami, N. Micali, and L. M. Scolaro, Kinetic effects of tartaric acid on the growth of chiral J-aggregates of tetrakis(4-sulfonatophenyl)porphyrin, *Chem. Commun.* **48**(40), 4872–4874 (2012).

62. M. A. Castriciano, A. Romeo, G. De Luca, V. Villari, L. M. Scolaro, and N. Micali, Scaling the chirality in porphyrin J-nanoaggregates, *J. Am. Chem. Soc.* **133**(4), 765–767 (2011).

63. R. Purrello, L. M. Scolaro, E. Bellacchio, S. Gurrieri, and A. Romeo, Chiral H- and J-type aggregates of meso-tetrakis(4-sulfonatophenyl)porphine on α-helical polyglutamic acid induced by cationic porphyrins, *Inorg. Chem.* **37**(14), 3647–3648 (1998).

64. L. Zhang, J. Yuan, and M. Liu, Supramolecular chirality of achiral TPPS complexed with chiral molecular films, *J. Phys. Chem. B* **107**(46), 12768–12773 (2003).

65. S. Jiang and M. Liu, Aggregation and induced chirality of an anionic meso-tetraphenylsulfonato porphyrin (TPPS) on a layer-by-layer assembled DNA/PAH matrix, *J. Phys. Chem. B* **108**(9), 2880–2884 (2004).

66. L. Zhang and M. Liu, Supramolecular chirality and chiral inversion of tetraphenylsulfonato porphyrin assemblies on optically active polylysine, *J. Phys. Chem. B* **113**(42), 14015–14020 (2009).

67. N. Berova and K. Nakanishi, Exciton chirality method: Principles and applications, in *Circular Dichroism*: *Principles and Applications*, N. Berova, K. Nakanishi, and R. W. Woody (Eds.), Wiley-VCH, New York, pp. 337–382 (2000).

68. C. Honda and H. Hada, Circular dichroism of polymolecular associate, J-aggregate, of 1,1'-2,2'-cyanine chloride by regular stirring solution, *Tetrahedron Lett.* **21**, 177–180 (1976).

69. B. Norden, Linear and circular dichroism of polymeric pseudocyanine, *J. Phys. Chem.* **81**(2), 157–159 (1977).

70. Z. El-Hachemi, O. Arteaga, A. Canillas, J. Crusats, J. Llorens, and J. M. Ribò, Chirality generated by flows in pseudocyanine dye J-aggregates: Revisiting 40 years old reports, *Chirality* **23**(8), 585–592 (2011).

71. O. Ohno, Y. Kaizu, and H. Kobayashi, J-aggregate formation of a water-soluble porphyrin in acidic aqueous media, *J. Chem. Phys.* **99**(5), 4128- (1993).

72. F. Hamba, K. Niimura, Y. Kitagawa, and K. Ishii, Helicity transfer in rotary evaporator flow, *Phys. Fluids* **26**, 017101 (2014).

73. R. Rubires, J.-A. Farrera, and J. M Ribó, Stirring effects on the spontaneous formation of chirality in the homoassociation of diprotonated meso-tetraphenylsulfonato porphyrins, *Chem. Eur. J.* **7**(2), 436–446 (2001).

74. C. Escudero, J. Crusats, I. Díez-Pérez, Z. El-Hachemi, and J. M. Ribó, Folding and hydrodynamic forces in J-aggregates of 5-phenyl-10, 15, 20-tris (4-sulfophenyl) porphyrin, *Angew. Chem. Int. Ed.* **45**(47), 8032–8035 (2006).

75. O. Arteaga, A. Canillas, J. Crusats, Z. El-Hachemi, J. Llorens, E. Sacristan, and J. M. Ribò, Emergence of supramolecular chirality by flows, *ChemPhysChem* **11**(16), 3511–3516 (2010).

76. A. Raudino and M. Pannuzzo, Hydrodynamic-induced enantiomeric enrichment of self-assemblies: Role of the solid-liquid interface in chiral nucleation and seeding, *J. Chem. Phys.* **137**(13), 134902 (2012).

77. J. M. Short, J. A. Berriman, C. Kubel, Z. El-Hachemi, J.-V. Naubron, and T. S. Balaban, Electron cryo-microscopy of TPPS$_4$·2HCl tubes reveals a helical organisation explaining the origin of their chirality, *ChemPhysChem* **14**(14), 3209–3214 (2013).

78. Z. El-Hachemi, C. Escudero, O. Arteaga, A. Canillas, J. Crusats, G. Mancini, R. Purrello, A. Sorrenti, A. D'urso, and J. M. Ribò, Chiral sign selection on the J-aggregates of diprotonated tetrakis-(4-sulfonatophenyl)porphyrin by traces of unidentified chiral contaminants present in the ultra-pure water used as solvent, *Chirality* **21**(4), 408–412 (2009).

79. N. Petit-Garrido, J. Claret, J. Ignés-Mullol, and F. Sagués, Stirring competes with chemical induction in chiral selection of soft matter aggregates, *Nat. Commun.* **3**, 1001 (2012).

80. G. P. Spada, Alignment by the convective and vortex flow of achiral self-assembled fibers induces strong circular dichroism effects, *Angew. Chem. Int. Ed.* **47**(4), 636–638 (2008).

81. A. Tsuda, Md. Akhtarul Alam, T. Harada, T. Yamaguchi, N. Ishii, and T. Aida, Spectroscopic visualization of vortex flows using dye-containing nanofibers, *Angew. Chem. Int. Ed.* **46**(43), 8198–8202 (2007).

82. M. Wolffs, S. J. George, Z. Tomovic, S. C. J. Meskers, A. P. H. J. Schenning, and E. W. Meijer, Macroscopic origin of circular dichroism effects by alignment of self-assembled fibers in solution, *Angew. Chem. Int. Ed.* **46**(43), 8203–8205 (2007).

83. A. Tsuda, Y. Nagamine, R. Watanabe, Y. Nagatani, N. Ishii, and T. Aida, Spectroscopic visualization of sound-induced liquid vibrations using a supramolecular nanofiber, *Nat. Chem.* **2**(11), 977–983 (2010).

84. O. Arteaga, A. Canillas, R. Purrello, and J. M. Ribó, Evidence of induced chirality in stirred solutions of supramolecular nanofibers, *Opt. Lett.* **34**(14), 2177–2179 (2009).

85. Z. El-Hachemi, O. Arteaga, A. Canillas, J. Crusats, C. Escudero, R. Kuroda, T. Harada, M. Rosa, and J. M. Ribó, On the mechano-chiral effect of vortical flows on the dichroic spectra of 5-phenyl-10,15,20-tris(4-sulfonatophenyl) porphyrin J-aggregates, *Chem. Eur. J.* **14**(21), 6438–6443 (2008).

86. P. W. G. Poodt, M. C. R. Heijna, P. C. M. Christianen, W. J. P. van Enckevort, W. J. de Grip, K. Tsukamoto, J. C. Maan, and E. Vlieg, Using gradient magnetic fields to suppress convection during crystal growth, *Cryst. Growth Des.* **6**(10), 2275–2280 (2006).

87. C. Papaseit, N. Pochon, and J. Tabony, Microtubule self-organization is gravity-dependent, *Proc. Natl. Acad. Sci. USA* **97**(15), 8364–8368 (2000).

88. L. D. Barron, Spin and gravity give a helping hand, *Nat. Chem.* **4**(3), 150–152 (2012).

89. L. D. Barron, False chirality, absolute enantioselection and CP violation: Pierre curie's legacy, *Magnetochemistry* **6**(1), 5 (2020).

90. S. Sevim, A. Sorrenti, C. Franco, S. Furukawa, S. Pané, A. J. deMello, and J. Puigmartí-Luis, Self-assembled materials and supramolecular chemistry within microfluidic environments: From common thermodynamic states to non-equilibrium structures, *Chem. Soc. Rev.* **47**(11), 3788–3803 (2018).

91. J. Puigmartí-Luis, Microfluidic platforms: A mainstream technology for the preparation of crystals, *Chem. Soc. Rev.* **43**(7), 2253–2271 (2014).

92. M. Numata and T. Kozawa, Two-Dimensional assembly based on flow supramolecular chemistry: Kinetic control of molecular interactions under solvent diffusion, *Chem. Eur. J.* **20**(21), 6234–6240 (2014).

93. A. Sorrenti, R. Rodriguez-Trujillo, D. B. Amabilino, and J. Puigmartí-Luis, Milliseconds make the difference in the far-from-equilibrium self-assembly of supramolecular chiral nanostructures, *J. Am. Chem. Soc.* **138**(22), 6920–6923 (2016).

94. B. L. Feringa and R. A. van Delden, Absolute asymmetric synthesis: The origin, control, and amplification of chirality, *Angew. Chem. Int. Ed.* **38**(23), 3418–3438 (1999).

95. M. P. Curie, Sur la symétrie dans les phénomènes physiques, symétrie d'un champ électrique et d'un champ magnétique, *J. Phys.* (Paris) **3**, 393–415 (1894).

96. M. Avalos, R. Babiano, P. Cintas, J. L. Jiménez, J. C. Palacios, and L. D. Barron, Absolute asymmetric synthesis under physical fields: Facts and fictions, *Chem. Rev.* **98**(7), 2391–2404 (1998).

97. A. Arlegui, B. Soler, A. Galindo, O. Arteaga, A. Canillas, J. M. Ribó, Z. El-Hachemi, J. Crusats, and A. Moyano, Spontaneous mirror-symmetry breaking coupled to top-bottom chirality transfer: From porphyrin self-assembly to scalemic Diels–Alder adducts, *Chem. Commun.* **55**(81), 12219–12222 (2019).

Index

www.ingramcontent.com/pod-product-compliance
Lightning Source LLC
Chambersburg PA
CBHW050559190326
41458CB00007B/2109